Notes on Estimation of Logistics Impact of Trade Policies

A Recursively Dynamic Applied General Equilibrium Approach

TURKAY YILDIZ

Copyright © 2015 Turkay Yildiz

All rights reserved.

ISBN-13: 978-1-329-88255-3

DEDICATION

To my parents...

TABLE OF CONTENTS

PREFACE x
Part I 12
Chapter 1. An overview of the computable general equilibrium (CGE) modeling and the *dynamic* CGE model 12
    1.1. History of CGE modeling...12
    1.2. The aim of CGE models and its mechanisms..............14
    1.3. The dynamic CGE model 16
    1.4. The GDyn 17
    References...21
Chapter 2. Theoretical structure of dynamic GTAP and the recursive dynamic model 24
    2.1. Theoretical structure of dynamic GTAP 24
    RunDynam and GEMPACK 27
    References...29
Chapter 3. Global Trade Analysis Project 32
The GTAP Modeling Framework 32
Recursive Dynamic GTAP...32
    Final demand: private consumption, saving and government 33
    GDyn 34
Part II 40
Chapter 4. The long-term assessment of exports of food and industrial products and its logistics impacts 40
(Simulation 1)..40
    1. Introduction..40
    2. Highlighted studies..41
    3. Framework, methodology, and results 45
    4. Conclusion 50
    References...51
Chapter 5. Assessment of exports of commodities and implication for logistics services 64
(Simulation 2)..64
    1. Introduction..64

- 2. Highlighted studies 65
- 3. Framework, methodology, and results 69
- 4. Conclusion 80
- References 81

Appendix A - Preliminaries 93
Preliminaries 102
- Files 102
- Sets 102
- "Read" Statements of Base Data 107
  - Saving 109
  - Government Consumption 110
  - Private Consumption 112
  - Firms 114
  - Global Trust 117
  - International Trade and Transport 118
  - Regional Household 120
- Common "Variables" 121
- Common "Coefficients" 126
  - Key Derivatives of the Base Data 126
  - Regional Expenditure and Income 130
  - Indirect Tax Receipts 132
  - Miscellaneous Coefficients 135

Appendix B - Modules 138
Modules 139
- 1. Government Consumption 140
  - 1-0. Module-Specific Variables 140
  - 1-1. Demands for Composite Goods 140
  - 1-2. Composite Tradeables 141
- 2. Private Consumption Module 143
  - 2-0. Module-Specific Variables 143
  - 2-1. Utility from Private Consumption 143
  - 2-2. Allen Partials, Price and Income Elasticities, Composite Demand 144
  - 2-3. Composite Tradeables 145
- 3. Firms 148
  - 3-0. Module-Specific Variables 149

- 3-1. Total Output Nest 150
- 3-2. Composite Intermediates Nest.................................. 151
- 3-3. Value-Added Nest 153
- 3-4. Zero Profits Equations.. 156
- 4. Physical Capital, Global Trust, and Savings 158
  - 4-0. Module-Specific Variables 159
  - 4-1. Equations of Notational Convenience 163
  - 4-2. Rate of Return Equations 165
  - 4-3. Required Rate of Growth in Rate of Return.......... 169
  - 4-4. Investment... 170
  - 4-5. Capital.. 172
  - 4-6. Global Trust 172
  - 4-7. Saving 173
- 5. International Trade.. 175
  - 5-1. Export Prices 175
  - 5-2. Demand for Imports 176
- 6. International Transport Services 180
  - 6-0. Module-Specific Variables and Coefficients 180
  - 6-1. Demand for Global Transport Services 183
  - 6-2. Supply of Transport Services 185
- 7. Regional Household... 189
  - 7-0. Module-Specific Coefficients................................... 189
  - 7-1. Supply of Endowments by the Regional Household 190
  - 7-2. Regional Wealth and Equity Income...................... 191
  - 7-3. Computation of Regional Income........................... 193
  - 7-4. Regional Household Demand System 196
  - 7-5. Aggregate Utility.. 197
- 8. Equilibrium Conditions... 200
  - 8-1. Market Clearing Conditions.................................... 200
  - 8-2. Net Foreign Assets and Liabilities.......................... 205
  - 8-3. Walras' Law 207
  - 8-4. Trade Balance Constraints 208

Appendices 211
Appendices 212
- A. Summary Indices... 212

   A-0. Appendix-Specific Variables and Coefficients.....212
   A-1. Factor Price Indices 213
   A-2. Regional Terms of Trade 215
   A-3. GDP Indices (Value, Price and Quantity)..............216
   A-4. Aggregate Trade Indices (Value, Price and Quantity) 219
   A-5. Trade Balance Indices 229
  B. Equivalent Variation..231
   B-0. Appendix-Specific Variables and Coefficients 231
   B-1. Government Consumption Shadow Demand System 233
   B-2. Private Consumption Shadow Demand System..234
   B-3. Regional Household Shadow Demand System....236
   B-4. Equivalent Variation..239
  C. Welfare Decomposition 240
  D. Terms of Trade Decomposition 261
ABOUT THE AUTHOR **269**

PREFACE

This book, is the result of my studies, which use the *dynamic* Global Trade Analysis Project (GTAP, GDyn) simulations for exploring impacts of international trade on logistics services. As the demand for logistics depends mostly on the volume of trade and trade patterns, international trade affects the transport and logistics, as it might generate a higher or lower demand for transport and logistics services in long-term.

This book consists of two parts and five chapters. First part of the book shortly introduces you to the general concepts of the computable general equilibrium models (CGE) and presents you fundamentals of a dynamic general equilibrium models. In each chapter of the last part, two short articles that simulate various scenarios are presented.

Each chapter of this book is independent of each other. I hope you will find this book informative, beneficial and appropriate for your needs.

Turkay Yildiz, M.A., Ph.D.

Part I

Chapter 1. An overview of the computable general equilibrium (CGE) modeling and the *dynamic* CGE model

This chapter provides you an overview on the basics of the computable general equilibrium modeling (CGE) and the dynamic CGE model. It first introduces you shortly to the historical roots of CGE modeling. It then introduces you to the aim and the main mechanisms of models. For static CGE modeling, refer to book and its chapters from Yildiz (2015).

1.1. History of CGE modeling

Walras (1874) formulated the first general equilibrium model, called the Walrasian model that could adapt to complex economic interactions. Based on the theoretical structure of the Walrasian general equilibrium, the CGE modeling approach is based on the work, which was completed in the 1950s by Kenneth Arrow and Gerard Debreu. CGE models are based on general equilibrium (GE) theories of Arrow and Debreu, where agents interact in competitive markets by introducing optimal prices and optimal amounts that satisfy markets and balance agents' equilibrium conditions.

Since then, CGE models have been developed in the early 1960s to solve for both market prices and quantities simultaneously, simulating the operation of a competitive market economy.

Finally, interest in these models has increased since their creation in the 1960s as applied economists have recognized the benefits of their use for the counterfactual analysis and improvement in the computations that allowed for detailed analysis. The first CGE model developed in Norway by Leif Johansen (Johansen, 1960), was designed to be used for policy analysis. The purpose of a CGE model is to try to model the entire economy and the relations between economic agents in it. Later, CGE models have been widely used to analyze the impact of macroeconomic policies and the impact of the allocation of development resources in developed countries since the early 1980s CGE models are preferred over the partial equilibrium models because they include the complex interdependencies in the analysis.

1.2. The aim of CGE models and its mechanisms

The aim of CGE models is to quantify the effects of the policy on equilibrium allocations and relative prices using the standard theory of general equilibrium.

CGE models have their roots in the input-output theory. They are widely used for economic, social and environmental planning, as it is effective for capturing inter-sectoral linkages. Indeed, CGE models are based on complex hypothetical mathematical relationships between different sectors of the economy that reflect the behavior of the key players in the economy and provide more realistic estimates than those obtained from input- models. However, input-output models have limitations, which simulated the development of CGE models. These limitations are reflected mainly in the assumptions of the model, including fixed price, the factor of the unlimited supply and fixed share factor and intermediate inputs in the production process. Under these assumptions, the input-output model cannot show substitution between production inputs, and reactive behavior of producers and consumers to changes in relative prices.

In short, CGE models are adapted to analyze contemporary strategic issues in a competitive market economy, as they have the price mechanism that plays an important role in the economy. The price mechanism is able to solve complex problems in an economy that cannot be achieved by input-output models. In general equilibrium models, economic agents make their own decisions on economic activities based on changes in the market price as indicated resource constraints and technology. In addition, the market balances supply and demand by adjusting prices. Because CGE models can quantify market behavior and changes, they are widely used in various policy analyzes. A CGE model describes an economy in equilibrium with prices and relative quantities, which are determined endogenously. While most empirical approaches to examine the impacts of policies and effects in a *ceteris paribus* condition, a CGE model, which provides comparative scenarios based on the baseline scenario incorporates the factor markets, markets for goods and foreign trade markets.

1.3. The dynamic CGE model

Analytical processing of overall economic growth has its origin in the work of early theorists such as Ramsey (1928), Solow (1956) and Koopmans (1965). However, because of their heavy computing requirements, real dynamic extensions of computable general equilibrium models are a recent development.

The extension of a static CGE model to a dynamic process is simple. Although computationally more complex, a dynamic CGE model differs from its static counterpart by including a driving force to move the economy from period to period. In most dynamic models, this force is provided by the growth of underlying labor and/or a change in the level of technology in one or more sectors of the economy. These changes are facilitated by new investments and growth in the capital stock in the economy.

As with the static model, real output for each sector in a particular reference year is reproduced through the calibration procedure. In addition, the economy is now expected to grow, and the initial benchmark must be run with all sectors, the quantities and production factors, each of which are needed to grow at the same steady state rate.

When counterfactual shock is given to a dynamic CGE model two things happen.

1. The affected prices and quantities traverse to a new growth path in the years following the shock.

2. The new growth path itself returns to a stable state, but with economic variables at a different level than they would have been in the benchmark case.

In general, the interest of these dynamic models is on this new path and how much higher or lower, it is than the initial benchmark path.

1.4. The GDyn

The Dynamic GTAP (Global Trade Analysis Project) model known as GDyn (see Ianchovichina and McDougall, 2012). Being a general equilibrium global trade model with dynamic elements (see Ianchovichina and McDougall, 2012 for details) the GDyn is an extension of the widely used GTAP model (Hertel, 1997).

It can be applied to analyze various issues, such

as trade policy, regional economic integration and climate change. The GTAP database model is a comparative static global model of general equilibrium, linking bilateral trade flows between all countries or regions, and explicitly models the consumption and production for all commodities of each national or regional economy.

As with other neoclassical CGE or applied general equilibrium (AGE) model in the GTAP, the producers are assumed to maximize profits and consumers are assumed to maximize utility (in GTAP). Product and factor market clearing requires that supply equals demand in each market (see Hertel, 1997 for details).

The GDyn is an extension of GTAP and retains its basic characteristics. It also provides an improved long-term treatment in the context of the GTAP data modeling.

It is a recursive model, generating a sequence of static equilibria based on the investment theory of adaptive expectations, and is bound by a number of dynamic characteristics.

The main features of the model include (Ianchovichina and McDougall, 2012, p 5.)

- the treatment of time; the distinction between physical and financial assets, and between domestic and foreign financial assets

- the treatment of capital and asset accumulation, assets and liabilities of firms and households, income from financial assets, and

- the investment theory of adaptive expectations

GTAP-Dyn is a recursive dynamic model applied general equilibrium (AGE) of the global economy. It extends the standard GTAP model (Hertel, 1997) to include:

- international capital mobility,
- capital accumulation
- adaptive expectations theory of investment.

Standard GTAP (Hertel and Tsigas, 1997) is a comparative static AGE model of the world economy, developed as a vehicle for teaching multi-country AGE modeling and to complement the GTAP multi-country AGE data base (Gehlhar, Gray, Hertel et al., 1997).

In general, it aims to provide a simple presentation of AGE modeling techniques widely used. It does, however, include some special features, notably an extensive decomposition of welfare results.

References

Yildiz, T. (2015) *Logistics Impact Assessment of Trade Policies by CGE Modeling: Theory and Practice.* ISBN: 978-1505609752

Yildiz, T. (2015). An overview of the computable general equilibrium (CGE) modeling. *Logistics Impact Assessment of Trade Policies by CGE Modeling: Theory and Practice*, 1 01/2015: chapter 1, ISBN: 978-1505609752

Yildiz, T. (2015). The standard CGE model with GAMS code. *Logistics Impact Assessment of Trade Policies by CGE Modeling: Theory and Practice*, 1 01/2015: chapter 2, ISBN: 978-1505609752

Yildiz, T. (2015). The economic rise of the Eurasian trade and its implications for logistics services in Turkey (Simulation 1). *Logistics Impact Assessment of Trade Policies by CGE Modeling: Theory and Practice*, 01/2015: chapter 3, ISBN: 978-1505609752

Yildiz, T. (2015). The Eurasian trade and the effects on logistics services in the EU: An assessment by using CGE modeling (Simulation 2). *Logistics Impact Assessment of Trade Policies by CGE Modeling: Theory and Practice*, 01/2015: chapter 4, ISBN: 978-1505609752

Yildiz, T. (2015). The rise of trade in NAFTA area and the logistics impact in the EU (Simulation 3). *Logistics Impact Assessment of Trade Policies by CGE Modeling: Theory and Practice*, 01/2015: chapter 5, ISBN: 978-1505609752

Yildiz, T. (2015). The economic rise of the Sub-Saharan Africa and the impact on logistics services in Turkey (Simulation 4). *Logistics Impact Assessment of Trade Policies by CGE Modeling: Theory and Practice*, 01/2015: chapter 6, ISBN: 978-1505609752

Yildiz, T. (2015). Rise of the trade between the Sub-Saharan Africa and the EU: Implications for logistics services (Simulation 5). *Logistics Impact Assessment of Trade Policies by CGE Modeling: Theory and Practice*, 01/2015: chapter 7, ISBN: 978-1505609752

Chapter 2. Theoretical structure of dynamic GTAP and the recursive dynamic model

The Dynamic GTAP model (GTAP-Dyn) is a recursively dynamic applied general equilibrium (AGE) model of the world economy. It extends the standard GTAP model (Hertel, 1997) to include:

- International capital mobility
- Capital accumulation
- An adaptive expectations theory of investment

2.1. Theoretical structure of dynamic GTAP

A salient technical feature of the new extension is the treatment of *time*. Many dynamic models treat time as an index, so that each of the variables in the model has a time index.

In GTAP-Dyn, time itself is a variable, subject to exogenous change with the usual policy, technology, and demographic variables.

The differences between standard GTAP and GTAP-Dyn model can be generalized as follows

(Ianchovichina and McDougall, 2001; Walmsley and Strutt, 2010):

1. Compared to the standard model of GTAP, GTAP-Dyn provides *a better long-term analysis*. Because the dynamic model needs to build the baseline scenario as well as take the accumulative effects of variable factors into consideration.

2. In the standard GTAP model, capitals are allowed to move between industries in a region, *but not between regions*. While in GTAP-Dyn, *capitals can move across regions*, which allow the investment allocation and endowment to respond to region-specific rates of return on capital.

3. The adjustment for the rate of return needs time. The standard GTAP model assumes that the rate of return adjustment in all countries is instantaneous without any delay. While in GTAP-Dyn we describe *a lagged adjustment*, which is more realistic.

4. GTAP-Dyn pulls in *the adaptive expectations theory of investment*. The investment movements depend on the changes of investors' expected

rates of rates other than the actual rates. Their expectations of rates of return may be in error in the short term, but remain consistent with long-term real rates.

5. GTAP-Dyn includes *the capitals and gains of financial assets* to achieve the dynamic links across years.

Dynamic general equilibrium models can be classified as truly dynamic (*"intertemporal"*) or sequential dynamic (*"recursive"*) models. Truly dynamic models are based on the theory of optimal growth where the behavior of economic agents is characterized by perfect foresights. They know everything about the future and react to future changes in prices.

- Households maximize their intertemporal utility function under a wealth constraint to determine their schedule of consumption over time.

- Investment decisions by firms are the result of the maximization of cash flow over the entire time horizon.

A recursive dynamic model is essentially a series of static CGE models, which are linked between periods by an exogenous and endogenous variable updating procedure.

The static CGE model used to develop a recursive dynamic process is based on several standard assumptions:

- constant returns to scale,
- perfect competition and price taking behavior,
- market-clearing conditions hold for commodities and primary factors,
- and zero profit conditions hold, implying that price equals marginal cost.

Recursive-dynamic models: multi-period CGE models in which results are computed one-period-at-a-time. In contrast, for intertemporal models, results are computed simultaneously for all periods.

Intertemporal models: multi-period CGE models in which results are computed simultaneously for all periods. In contrast, for recursive-dynamic models, results are computed one-period-at-a-time.

RunDynam and GEMPACK

RunDynam allows you to build a reference scenario (which may be a forecast) and policy deviations from the base case with a model, which was implemented using GEMPACK. The model is solved on a year to year basis (that is, recursively) over a number of years, from initial data. For each year thereafter, the input data is the data updated by the previous simulation.

1. First, you solve the base case, and then carry out the policy deviation. You can choose a group of input data files for the model; these are the starting points for your base case.

2. You specify the closures and the impact on text files, using the syntax required in GEMPACK command files.

3. You can choose names for the output files of the base case and of policy runs.

4. You can choose from several methods to solve the model.

5. You can view the results of the base case or the deviation of the policy on the screen or export them to other programs.

6. RunDynam can produce graphics of selected variables over time.

7. You can view or copy either the initial model database or any of the updated data files produced during the base case or policy deviation.

References

Farmer, K., & Wendner, R. (2004). Dynamic multi-sector CGE modeling and the specification of capital. *Structural Change and Economic Dynamics*, 15(4), 469-492. doi: http://dx.doi.org/10.1016/j.strueco.2003.12.002

Auerbach, A.J., Kotlikoff, L.J., 1987. *Dynamic Fiscal Policy*. Cambridge University Press, Cambridge.

P.A. Diamond, National debt in a neoclassical growth model, *American Economic Review*, 55 (1965), pp. 1135–1150

Chapter 3. Global Trade Analysis Project

The GTAP Modeling Framework

Recursive Dynamic GTAP

The main objective of GTAP-Dyn is to provide better long-term treatment in the GTAP framework. In standard GTAP, capital can move between industries in a region, *but not between regions*. This hinders analysis of policy shocks and other developments diversely affecting incentives to invest in different regions. For a good long-term treatment, we need *international capital mobility*.

The main distinctive features of GDyn are its specification investment income flows associated with financial assets. The model distinguishes between *physical* and *financial* assets, and in the latter between *domestic* and *foreign*.

The model allows to determine the accumulation of capital and assets of each national economy, and the assets and liabilities of businesses and households in each region. The theory of investment in each region is characterized by *adaptive expectations*, in which the differences between *actual* and *expected rates of return* are corrected over time by displacing investment and international capital mobility.

The GDyn uses a simplified and unified treatment of the mobility of capital and investment in the context of a global CGE model. This specification endogenously captures the overall effects of the accumulation of capital and wealth in the country, and the effects of income from foreign ownership of assets.

Final demand: private consumption, saving and government

Final demand in each region is represented by an aggregate called "*Regional Household*," which is a Cobb-Douglas combination of private household consumption, saving and government spending.

Private consumption optimizer is represented by an agent governed by a function of spending CDE (constant difference of elasticity).

Government consumption follows a Cobb–Douglas function, which implies a constant share of public spending on goods and services. The savings is a residual element of the country's income and determines the net investment in the economy.

GDyn

The standard version of GDyn is a recursive-dynamic extension of the standard GTAP (Hertel, 1997), developed for better treatment of medium and long-term simulations, as it strengthens the investment side of the modeling framework to enable international capital mobility (Ianchovichina & Walmsley, 2012).

GDyn extends the standard, comparative static version of the GTAP model by introducing

- international capital mobility,
- endogenous capital accumulation and
- adaptive expectations theory of investment in a recursive dynamics setting.

GDyn is a real assets model, i.e. investment is associated with equity: the regional households (shareholders) own equity in the firm equal to the value of physical capital and earn income (dividends) corresponding to their ownership share - there are no financial markets and no differentiation between debt and equity.

The model keeps track of gross ownership positions and income flows associated with them and thus compared to the comparative static version the GTAP model is augmented to improve the representation of balance of payments relationships.

Despite the advantages offered by perfect foresight models, the solution procedure chosen for the GDyn is a recursive one in which investors are allowed to have errors in their expectations, i.e. a novel adaptive expectations specification of investors' behavior. Compared to perfect foresight models GDyn offers greater empirical realism, flexibility in data specification and lower computational complexity.

GDyn inherits the treatment of savings of the comparative static GTAP model. As implemented by, the representative household allocates regional income that would maximize the per capita utility based on a Cobb–Douglas utility function complemented with non-homothetic preferences on the private consumption side. Real saving is a single commodity that is defined as savings deflated by the price of savings. The Cobb–Douglas specification keeps the budget shares constant, implicitly assuming a constant marginal propensity to save of the household.

Capital goods are a production sector and their supply is determined by a Leontieff type production technology. On the other hand, capital is a value added component and is a direct input into production of all goods (except capital goods) governed by a CES type allocation. Capital is assumed to be perfectly mobile across sectors determining a single rental rate across sectors that clears the market.

As in most recursive dynamic models, each period's equilibrium determines the level of global savings and implicitly the aggregate amount of investment expenditure available in that specific period. International capital mobility is modeled using a disequilibrium approach that reconciles investment theory with empirical findings.

The disequilibrium approach adopted here is described by two mechanisms in the model:

1. There is a gradual convergence of the expected rate of return leading to the equalization of expected rates of return on the long run; and

2. Errors in expectations with respect to the actual rate of return are eliminated over time.

Investors are assumed to respond to expected rates of return as opposed to actual rates of return when making investment decisions allowing for errors in expectations. For instance, when investment in the base data is low despite high actual rates of return it is assumed to be due to errors in expectations; investors are assumed to behave adaptively and over time these errors are eliminated and the expected rate of return will converge toward the observed rate of return.

The GDyn model in its current form does not make use of portfolio allocation theory in determining gross ownership positions, i.e. investors reactions are based only on (expected) rates of return and hence the GDyn model is an investment demand driven model.

Moreover, domestic firms hold equity directly in domestic firms, the lack of availability of bilateral data on foreign assets, precludes the representative household from holding equity directly in foreign firms. This lack of bilateral data on foreign assets and liabilities compels many CGE modelers to employ a somewhat artificial representation of foreign investment. The GDyn model overcomes this problem through the adoption of a fictional entity called the global trust. The global trust collects the saving of all the regional households and allocates this to regional investment on their behalf.

The mechanism that the GDyn model uses to determine the composition of the cross-ownership matrix over time is cross-entropy minimization. The choice of the cross-entropy allocation of wealth is motivated by the fact that this type of specification is able to reproduce some of the empirical findings of the investment literature such as the home bias of puzzle of investment.

Part II

Chapter 4. The long-term assessment of exports of food and industrial products and its logistics impacts (Simulation 1)

1. Introduction

This study examines the fictitious long-term trade among North America (NAM), European Union (EUN) and the rest of the world (ROW) and it effects on the logistics services (food, mnfc, serv). Based on the scenario, the results indicate that various logistics services to/from economies exhibit increases or decreases. Results are reported using *dynamic* computable general equilibrium (CGE) simulations.

The remainder of this study is organized as follows. Section 2 presents some highlighted studies. Section 3 presents the framework, methodology, results and short discussion. The study is concluded in Section 4.

2. Highlighted studies

Feraboli (2007) studied preferential trade liberalization, fiscal policy responses and welfare for Jordan. Lucke *et al.* (2007) assessed economic and fiscal reforms in Lebanon with debt constraints. Radulescu and Stimmelmayr (2010) explored the impact of the 2008 German corporate tax reform. Lu *et al.* (2010) studied the impacts of carbon tax and complementary policies on Chinese economy. Fehr (2000) analyzed consumption taxation. Cho *et al.* (2010) researched allocation and banking in Korean permits trading.

Espinosa *et al.* (2014) performed ex-ante analysis of the Regional Impacts of the common agricultural policy. Xu *et al.* (2011) explored impacts of agricultural public spending on Chinese food economy. Dogruel *et al.* (2003) researched macroeconomics of Turkey's agricultural reforms. Femenia (2010) investigated impacts of stockholding behavior on agricultural market volatility.

Bruvoll and Foehn (2006) analyzed trans-boundary effects of environmental policy. Bretschger *et al.*

(2011) investigated growth effects of carbon policies. Chi *et al.* (2014) performed scenarios analysis of the energies' consumption and carbon emissions in China. Doumax *et al.* (2014) investigated biofuels, tax policies and oil prices in France. He *et al.* (2014) performed low-carbon-oriented dynamic optimization of residential energy pricing for China. Takeda (2007) investigated the double dividend from carbon regulations in Japan. Rive *et al.* (2006) investigated climate agreements based on responsibility for global warming. Schenker (2013) investigated exchanging goods and damages from the perspective of the role of trade on the distribution of climate change costs. Markandya *et al.* (2015) analyzed trade-offs in international climate policy options. Mori *et al.* (2006) studied integrated assessments of global warming issues. Okagawa *et al.* (2012) assessed GHG emission reduction pathways in a society without carbon capture and nuclear technologies. O'Ryan *et al.* (2011) studied the socioeconomic and environmental effects of free trade agreements for Chile. Loisel (2009) explored environmental climate instruments in Romania. Kishimoto *et al.* (2014) modeled regional transportation demand in China and explored the impacts of a national carbon policy. Fujino *et al.* (2006) performed multi-gas mitigation analysis on stabilization scenarios using aim global model. Xie *et al.* (2015) studied disaster risk decision of regional mitigation Investment.

Georges (2008) analyzed liberalizing NAFTA Rules of Origin. Vellinga (2008) commented on dynamic general-equilibrium model of an open economy. Brocker and Korzhenevych (2013) explored forward-looking dynamics in spatial CGE modelling. Giesecke (2002) explained regional economic performance. Deepak *et al.* (2001) studied local government portfolios and regional growth. Hubler (2011) investigated technology diffusion under contraction and convergence of China.

Zhang (2001) performed iterative method for finding the balanced growth solution of the non-linear dynamic input-output model. Kristkova (2012) explored impact of R&D investment on economic growth of the Czech Republic. Lay *et al.* (2008) studied shocks, policy reforms and pro-poor growth in Bolivia. Ihori *et al.* (2011) studied health insurance reform and economic growth for Japan. Breisinger *et al.* (2011) explored impacts of the triple global crisis on growth and poverty for Yemen. Breisinger *et al.* (2009) modeled growth options and structural change to reach middle-income country status for Ghana.

Wu and Xiao (2014) run dynamic CGE model and performed simulation analysis on the impact of citizenization of Rural migrant workers on the labor and capital Markets in China. AlShehabi (2013) modeled energy and labor linkages for Iran.

Loisel (2010) explored quota allocation rules in Romania which are assessed by a dynamic CGE model. Wittwer (2009) explored the economic impacts of a new dam in South-East Queensland.

Ozdemir and Bayar (2009) analyzed the peace dividend effect of Turkish convergence to the EU using a multi-region dynamic CGE model for Greece and Turkey. Aydin and Acar (2011) studied economic impact of oil price shocks on the Turkish economy in the coming decades. Barkhordar and Saboohi (2013) assessed alternative options for allocating oil revenue in Iran.

Dixon *et al.* (2011) explored the economic costs to the U.S. of closing its borders. Cardenete and Delgado (2015) simulated the impact of withdrawal of European funds on Andalusian economy.

Fougere *et al.* (2007) performed a sectoral and occupational analysis on ageing population in Canada. Dixon and Rimmer (2010) validated a detailed, dynamic CGE Model of the USA. Dixon *et al.* (2005) studied rational expectations for large CGE models.

3. Framework, methodology, and results

The GDyn is an extension of GTAP and retains its basic characteristics. It also provides an improved long-term treatment in the context of the GTAP data modeling. It is a recursive model, generating a sequence of static equilibria based on the investment theory of adaptive expectations, and is bound by a number of dynamic characteristics. GTAP-Dyn provides a better long-term analysis. A recursive dynamic model is essentially a series of static CGE models, which are linked between periods by an exogenous and endogenous variable updating procedure.

This study examines the fictitious long-term trade among North America (NAM), European Union (EUN) and the rest of the world (ROW) and it effects on the logistics services. This study looks into the effects on logistics services based on the scenario of an economy wide technology shock in ROW economies. The results indicate that various logistics services from/to economies exhibit increases or decreases (See Tables 1 through 4).

Policy Shock

ashock afereg("*row*") = -5 ;

afereg (REG): Economy wide afe shock

afe (ENDW_COMM, PROD_COMM, REG) : Primary factor i augmenting tech change by j of r

Table 1. Food, Mnfc, Serv 2007 (percentage changes)

qxs[food**](D)	1 NAM	2 EUN	3 ROW
1 NAM	-0.013	-0.693	-2.309
2 EUN	1.231	0.440	-1.186
3 ROW	-0.212	-0.952	-2.592

qxs[mnfc**](D)	1 NAM	2 EUN	3 ROW
1 NAM	-0.444	-1.467	-6.705
2 EUN	1.365	0.283	-5.068
3 ROW	3.575	2.441	-3.044

qxs[serv**](D)	1 NAM	2 EUN	3 ROW
1 NAM	-0.647	-1.066	-5.821
2 EUN	0.702	0.304	-4.523
3 ROW	2.286	1.899	-3.035

Table 2. Food, Mnfc, Serv 2012 (percentage changes)

qxs[food**](D)	1 NAM	2 EUN	3 ROW
1 NAM	0.926	0.723	0.393
2 EUN	1.640	1.509	1.128
3 ROW	-3.950	-4.098	-4.425

qxs[mnfc**](D)	1 NAM	2 EUN	3 ROW
1 NAM	1.735	0.998	-2.829
2 EUN	2.622	1.924	-1.975
3 ROW	-1.027	-1.764	-5.478

qxs[serv**](D)	1 NAM	2 EUN	3 ROW
1 NAM	0.848	0.374	-2.995
2 EUN	1.890	1.437	-1.970
3 ROW	-1.535	-1.957	-5.282

Table 3. Food, Mnfc, Serv 2017 (percentage changes)

qxs[food**](D)	1 NAM	2 EUN	3 ROW
1 NAM	1.578	1.773	2.558
2 EUN	1.829	2.226	2.923
3 ROW	-6.683	-6.385	-5.665

qxs[mnfc**](D)	1 NAM	2 EUN	3 ROW
1 NAM	2.908	2.504	0.599
2 EUN	2.973	2.685	0.719
3 ROW	-4.738	-5.092	-6.811

qxs[serv**](D)	1 NAM	2 EUN	3 ROW
1 NAM	1.810	1.272	-0.565
2 EUN	2.417	1.902	0.050
3 ROW	-4.389	-4.854	-6.615

Table 4. Food, Mnfc, Serv 2020 (percentage changes)

qxs[food**](D)	1 NAM	2 EUN	3 ROW
1 NAM	1.723	2.104	3.049
2 EUN	1.705	2.311	3.165
3 ROW	-7.238	-6.757	-5.892

qxs[mnfc**](D)	1 NAM	2 EUN	3 ROW
1 NAM	3.087	2.883	1.516
2 EUN	2.754	2.682	1.260
3 ROW	-5.581	-5.738	-6.940

qxs[serv**](D)	1 NAM	2 EUN	3 ROW
1 NAM	2.010	1.521	0.033
2 EUN	2.305	1.840	0.345
3 ROW	-4.971	-5.388	-6.808

Commodity	Logistics Service
Sea transport services	Maritime Transport
Air transport services	Air Transport
Oil, gas, petroleum, coal products, chemicals, rubber, plastic products	Liquid Bulk Products
Motor vehicles and parts	Special
Live animals	Break Bulk Products
Paddy rice, vegetables, fruits, animal products, raw milk, wool, silk-worm cocoons, fishery, cattle meat, sheep meat, horse meat, meat products nec, vegetable oils and fats, dairy products, processed rice, food products nec.	Containerazable Agricultural Products
Beverages, tobacco products, textiles, wearing apparel, leather products, wood products, paper products, metals nec, metal products, transport equipment nec, electronic equipment, machinery and equipment nec, manufactures nec.	Containerazable Manufactured Products
Wheat, cereal grains nec, oil seeds, sugar cane, sugar beet, plant-based fibers, crops nec, forestry, coal, minerals nec, sugar, mineral products nec, ferrous metals.	Dry Bulk Products
Electricity, gas and water, construction, trade, transport nec, communication, financial services nec, insurance, business services nec, recreational services, public administration and defense, health, education and dwellings	Other Services

Figure 1. Commodities and sample aggregation (Source: GTAP Database)

Figure 1 depicts a sample aggregation of the commodities to represent their corresponding logistics services.

4. Conclusion

This study looked into the effects on logistics services based on the scenario of an economy wide technology shock in ROW economies. The results indicate that various logistics services from/to economies exhibit increases or decreases. These results are reported by using dynamic (CGE) simulations.

References

AlShehabi, O. H. (2013). Modelling energy and labour linkages: A CGE approach with an application to Iran. *Economic Modelling, 35*, 88-98. doi: 10.1016/j.econmod.2013.06.047

Aydin, L., & Acar, M. (2011). Economic impact of oil price shocks on the Turkish economy in the coming decades: A dynamic CGE analysis. *Energy Policy, 39*(3), 1722-1731. doi: 10.1016/j.enpol.2010.12.051

Barkhordar, Z. A., & Saboohi, Y. (2013). Assessing alternative options for allocating oil revenue in Iran. *Energy Policy, 63*, 1207-1216. doi: 10.1016/j.enpol.2013.08.099

Breisinger, C., Diao, X. S., Collion, M. H., & Rondot, P. (2011). Impacts of the Triple Global Crisis on Growth and Poverty: The Case of Yemen. *Development Policy Review, 29*(2), 155-184. doi: 10.1111/j.1467-7679.2011.00530.x

Breisinger, C., Diao, X. S., & Thurlow, J. (2009). Modeling growth options and structural change to reach middle income country status: The case of Ghana. *Economic Modelling, 26*(2), 514-525. doi: 10.1016/j.econmod.2008.10.007

Bretschger, L., Ramer, R., & Schwark, F. (2011). Growth effects of carbon policies: Applying a fully dynamic CGE model with heterogeneous capital. *Resource and Energy Economics, 33*(4), 963-980. doi: 10.1016/j.reseneeco.2011.06.004

Brocker, J., & Korzhenevych, A. (2013). Forward looking dynamics in spatial CGE modelling. *Economic Modelling, 31,* 389-400. doi: 10.1016/j.econmod.2012.11.031

Bruvoll, A., & Foehn, T. (2006). Transboundary effects of environmental policy: Markets and emission leakages. *Ecological Economics, 59*(4), 499-510. doi: 10.1016/j.ecolecon.2005.11.015

Cardenete, M. A., & Delgado, M. C. (2015). A simulation of impact of withdrawal European funds on Andalusian economy using a dynamic CGE model: 2014-20. *Economic Modelling, 45,* 83-92. doi: 10.1016/j.econmod.2014.09.021

Chi, Y. Y., Guo, Z. Q., Zheng, Y. H., & Zhang, X. P. (2014). Scenarios Analysis of the Energies' Consumption and Carbon Emissions in China Based on a Dynamic CGE Model. *Sustainability, 6*(2), 487-512. doi: 10.3390/su6020487

Cho, G. L., Kim, H. S., & Kim, Y. D. (2010). Allocation and banking in Korean permits trading. *Resources Policy, 35*(1), 36-46. doi: 10.1016/j.resourpol.2009.10.001

Deepak, M. S., West, C. T., & Spreen, T. H. (2001). Local government portfolios and regional growth: Some combined dynamic CGE/optimal control results. *Journal of Regional Science, 41*(2), 219-254. doi: 10.1111/0022-4146.00215

Dixon, P. B., Giesecke, J. A., Rimmer, M. T., & Rose, A. (2011). The economic costs to the U.S. Of closing its borders: a computable general equilibrium analysis. *Defence and Peace Economics, 22*(1), 85-97. doi: 10.1080/10242694.2010.491658

Dixon, P. B., Pearson, K. R., Picton, M. R., & Rimmer, M. T. (2005). Rational expectations for large CGE models: A practical algorithm and a policy application. *Economic Modelling, 22*(6), 1001-1019. doi: 10.1016/j.econmod.2005.06.007

Dixon, P. B., & Rimmer, M. T. (2010). Validating a Detailed, Dynamic CGE Model of the USA*.

Economic Record, 86, 22-34. doi: 10.1111/j.1475-4932.2010.00656.x

Dogruel, F., Dogruel, A. S., & Yeldan, E. (2003). Macroeconomics of Turkey's agricultural reforms: an intertemporal computable general equilibrium analysis. *Journal of Policy Modeling, 25*(6-7), 617-637. doi: 10.1016/s0161-8938(03)00056-5

Doumax, V., Philip, J. M., & Sarasa, C. (2014). Biofuels, tax policies and oil prices in France: Insights from a dynamic CGE model. *Energy Policy, 66*, 603-614. doi: 10.1016/j.enpol.2013.11.027

Espinosa, M., Psaltopoulos, D., Santini, F., Phimister, E., Roberts, D., Mary, S., . . . Paloma, S. G. Y. (2014). Ex-Ante Analysis of the Regional Impacts of the Common Agricultural Policy: A Rural-Urban Recursive Dynamic CGE Model Approach. *European Planning Studies, 22*(7), 1342-1367. doi: 10.1080/09654313.2013.786683

Fehr, H. (2000). From destination- to origin-based consumption taxation: A dynamic CGE analysis. *International Tax and Public Finance, 7*(1), 43-61. doi: 10.1023/a:1008754029145

Femenia, F. (2010). Impacts of Stockholding Behaviour on Agricultural Market Volatility: A Dynamic Computable General Equilibrium Approach. *German Journal of Agricultural Economics, 59*(3), 187-201.

Feraboli, O. (2007). Preferential trade liberalisation, fiscal policy responses and welfare: A dynamic CGE model for Jordan. *Jahrbucher Fur Nationalokonomie Und Statistik, 227*(4), 335-357.

Fougere, M., Mercenier, J., & Merette, M. (2007). A sectoral and occupational analysis of population ageing in Canada using a dynamic CGE overlapping generations model. *Economic Modelling, 24*(4), 690-711. doi: 10.1016/j.econmod.2007.01.001

Fujino, J., Nair, R., Kainuma, M., Masui, T., & Matsuoka, Y. (2006). Multi-gas mitigation analysis on stabilization scenarios using aim global model. *Energy Journal*, 343-353.

Georges, P. (2008). Liberalizing NAFTA Rules of Origin: A Dynamic CGE Analysis. *Review of*

International Economics, 16(4), 672-691. doi: 10.1111/j.1467-9396.2008.00771.x

Giesecke, J. (2002). Explaining regional economic performance: An historical application of a dynamic multi-regional CGE model. *Papers in Regional Science, 81*(2), 247-278. doi: 10.1007/s101100100100

He, Y. X., Liu, Y. Y., Wang, J. H., Xia, T., & Zhao, Y. S. (2014). Low-carbon-oriented dynamic optimization of residential energy pricing in China. *Energy, 66,* 610-623. doi: 10.1016/j.energy.2014.01.051

Hubler, M. (2011). Technology diffusion under contraction and convergence: A CGE analysis of China. *Energy Economics, 33*(1), 131-142. doi: 10.1016/j.eneco.2010.09.002

Ihori, T., Kato, R. R., Kawade, M., & Bessho, S. (2011). Health insurance reform and economic growth: Simulation analysis in Japan. *Japan and the World Economy, 23*(4), 227-239. doi: 10.1016/j.japwor.2011.07.003

Kishimoto, P. N., Zhang, D., Zhang, X. L., & Karplus, V. J. (2014). Modeling Regional Transportation Demand in China and the

Impacts of a National Carbon Policy. *Transportation Research Record*(2454), 1-11. doi: 10.3141/2454-01

Kristkova, Z. (2012). Impact of R&D Investment on Economic Growth of The Czech Republic - A Recursively Dynamic CGE Approach. *Prague Economic Papers, 21*(4), 412-433.

Lay, J., Thiele, R., & Wiebelt, M. (2008). Shocks, policy reforms and pro-poor growth in Bolivia: A simulation analysis. *Review of Development Economics, 12*(1), 37-56. doi: 10.1111/j.1467-9361.2007.00394.x

Loisel, R. (2009). Environmental climate instruments in Romania: A comparative approach using dynamic CGE modelling. *Energy Policy, 37*(6), 2190-2204. doi: 10.1016/j.enpol.2009.02.001

Loisel, R. (2010). Quota allocation rules in Romania assessed by a dynamic CGE model. *Climate Policy, 10*(1), 87-102. doi: 10.3763/cpol.2008.0557

Lu, C. Y., Tong, Q., & Liu, X. M. (2010). The impacts of carbon tax and complementary policies on Chinese economy. *Energy Policy,*

38(11), 7278-7285. doi: 10.1016/j.enpol.2010.07.055

Lucke, B., Soto, B. G., & Zotti, J. (2007). Assessing economic and fiscal reforms in Lebanon - A dynamic CGE analysis with debt constraints. *Emerging Markets Finance and Trade, 43*(1), 35-63. doi: 10.2753/ree1540-496x430102

Markandya, A., Antimiani, A., Costantini, V., Martini, C., Palma, A., & Tommasino, M. C. (2015). Analyzing Trade-offs in International Climate Policy Options: The Case of the Green Climate Fund. *World Development, 74,* 93-107. doi: 10.1016/j.worlddev.2015.04.013

Mori, S., Akimoto, K., Homma, T., Sano, F., Oda, J., Hayashi, A., . . . Tomoda, T. (2006). Integrated assessments of global warming issues and an overview of project PHOENIX - A comprehensive approach. *Ieej Transactions on Electrical and Electronic Engineering, 1*(4), 383-396. doi: 10.1002/tee.20081

Okagawa, A., Masui, T., Akashi, O., Hijioka, Y., Matsumoto, K., & Kainuma, M. (2012). Assessment of GHG emission reduction pathways in a society without carbon capture and nuclear technologies. *Energy Economics, 34*, S391-S398. doi: 10.1016/j.eneco.2012.07.011

O'Ryan, R., De Miguel, C. J., Miller, S., & Pereira, M. (2011). The Socioeconomic and environmental effects of free trade agreements: a dynamic CGE analysis for Chile. *Environment and Development Economics, 16*, 305-327. doi: 10.1017/s1355770x10000227

Ozdemir, D., & Bayar, A. (2009). The Peace Dividend Effect of Turkish Convergence to The EU: A Multi-Region Dynamic CGE Model Analysis For Greece And Turkey. *Defence and Peace Economics, 20*(1), 69-78. doi: 10.1080/10242690701833217

Radulescu, D., & Stimmelmayr, M. (2010). The impact of the 2008 German corporate tax reform: A dynamic CGE analysis. *Economic Modelling, 27*(1), 454-467. doi: 10.1016/j.econmod.2009.10.012

Rive, N., Torvanger, A., & Fuglestvedt, J. S. (2006). Climate agreements based on responsibility for global warming: Periodic updating, policy choices, and regional costs. *Global Environmental Change-Human and Policy*

Dimensions, 16(2), 182-194. doi: 10.1016/j.gloenvcha.2006.01.002

Schenker, O. (2013). Exchanging Goods and Damages: The Role of Trade on the Distribution of Climate Change Costs. *Environmental & Resource Economics, 54*(2), 261-282. doi: 10.1007/s10640-012-9593-z

Takeda, S. (2007). The double dividend from carbon regulations in Japan. *Journal of the Japanese and International Economies, 21*(3), 336-364. doi: 10.1016/j.jjie.2006.01.002

Vellinga, N. (2008). Dynamic general-equilibrium model of an open economy: A comment. *Journal of Policy Modeling, 30*(6), 993-997. doi: 10.1016/j.jpolmod.2007.04.009

Wittwer, G. (2009). The Economic Impacts of a New Dam in South-East Queensland. *Australian Economic Review, 42*(1), 12-23. doi: 10.1111/j.1467-8462.2009.00506.x

Wu, Q., & Xiao, H. (2014). Dynamic CGE Model and Simulation Analysis on the Impact of Citizenization of Rural Migrant Workers on the Labor and Capital Markets in China.

Discrete Dynamics in Nature and Society, 8. doi: 10.1155/2014/351947

Xie, W., Li, N., Wu, J. D., & Hao, X. L. (2015). Disaster Risk Decision: A Dynamic Computable General Equilibrium Analysis of Regional Mitigation Investment. *Human and Ecological Risk Assessment, 21*(1), 81-99. doi: 10.1080/10807039.2013.871997

Xu, S. W., Zhang, Y. M., Diao, X. S., & Chen, K. Z. (2011). Impacts of agricultural public spending on Chinese food economy A general equilibrium approach. *China Agricultural Economic Review, 3*(4), 518-534. doi: 10.1108/17561371111192365

Zhang, J. S. (2001). Iterative method for finding the balanced growth solution of the non-linear dynamic input-output model and the dynamic CGE model. *Economic Modelling, 18*(1), 117-132. doi: 10.1016/s0264-9993(00)00031-6

Chapter 5. Assessment of exports of commodities and implication for logistics services (Simulation 2)

1. Introduction

In this study, a fictitious trade and its 3-year-term potential effects on logistics services are simulated. The results indicate that some logistics services (food, mnfcs, svces) are much more sensitive to changes in some economies. These results are simulated and reported by using *dynamic* general equilibrium models. Based on the scenario, the results indicate that various logistics services to/from economies exhibit increases or decreases.

The remainder of this study is organized as follows. Section 2 presents some highlighted studies. Section 3 presents the framework, methodology, results and short discussion. The study is concluded in Section 4.

2. Highlighted studies

Parrado and De Cian (2014) explored technology spillovers embodied in international trade. Philip *et al.* (2014) investigated technological change in irrigated agriculture in a semiarid region of Spain.

Schenker (2013) investigated exchanging goods and damages as a role of trade on the distribution of climate change costs. Qin *et al.* (2011) assessed economic impacts of China's water-pollution-mitigation-measures through a dynamic computable general equilibrium analysis.

Xie *et al.* (2015) studied disaster risk decision by analyzing regional mitigation investment. Xie *et al.* (2014) modeled the economic costs of disasters and recovery. Gohin and Rault (2013) assessed the economic costs of a foot and mouth disease outbreak on Brittany by using a dynamic computable general equilibrium.

Berrittella and Zhang (2015) investigated fiscal sustainability in the EU. Bhattarai and Dixon (2014) explored equilibrium unemployment in a general equilibrium model with taxes. Mabugu *et al.* (2015) analyzed pro-poor tax policy changes in South Africa and investigated potentials and limitations.

Bhattarai (2015) investigated financial deepening and economic growth. Cheong and Tongzon (2013) compared the economic impact of the Trans-Pacific Partnership and the regional comprehensive economic partnership. Decreux and Fontagne (2015) investigated multilateral trade talks. Jiang and Mai (2015) studied social welfare housing project and its effects in China. Matovu (2012) investigated trade reforms and horizontal inequalities for Uganda. Seung and Kraybill (2001) explored the effects of infrastructure investment for Ohio. Verikios *et al.* (2015) studied improving health in an advanced economy for Australia. Lakatos and Walmsley (2012) studied investment creation and diversion effects of the ASEAN-China free trade agreement. Boccanfuso *et al.* (2014) performed a comparative analysis of funding schemes for public infrastructure spending in Quebec.

Barkhordar and Saboohi (2013) assessed alternative options for allocating oil revenue in Iran. Breisinger *et al.* (2012) investigated leveraging fuel subsidy reform for transition in Yemen. Dai *et al.* (2012) explored the impacts of China's household consumption expenditure patterns on energy demand and carbon emissions towards 2050. Faehn and Bruvoll (2009) investigated richer and cleaner issues at others expense. Hosoe (2014) investigated Japanese manufacturing facing post-Fukushima power crisis. Liang *et al.* (2014) studied platform for China energy & environmental policy analysis. Ruamsuke *et al.* (2015) explored energy and economic impacts of the global climate change policy on Southeast Asian countries. Wu *et al.* (2014) analyzed the future vehicle-energy-demand in China.

Femenia (2010) explored impacts of stockholding behaviour on agricultural market volatility. Femenia and Gohin (2013) studied optimal implementation of agricultural policy reforms. Furuya *et al.* (2015) studied economic evaluation of agricultural mitigation and adaptation technologies for Climate Change. Mariano *et al.* (2015) studied the effects of domestic rice market interventions outside business-as-usual conditions for imported rice prices. Akune *et al.* (2015) studied economic evaluation of dissemination of high temperature-tolerant rice in Japan. Bourne *et al.* (2012) measured the impact of the global financial crisis on Spanish agriculture. Breisinger and Ecker (2014) simulated economic growth effects on food and nutrition security in Yemen. Philippidis and Hubbard (2005) wrote a note on the ban of UK beef exports.

Arndt *et al.* (2012) studied biofuels and economic development for Tanzania. Asafu-Adjaye and Mahadevan (2013) studied implications of CO_2 reduction policies for a high carbon emitting economy. Bao *et al.* (2013) studied impacts of border carbon adjustments on China's sectoral emissions. Chi *et al.* (2014) performed scenarios analysis of the energies' Consumption and carbon emissions in China. Doumax *et al.* (2014) studied biofuels, tax policies and oil prices in France. Lanzi *et al.* (2012) studied alternative approaches for levelling carbon prices in a world with fragmented carbon markets. Liang *et al.* (2013) assessed the distributional impacts of carbon tax among households across different income groups of China. Liu and Lu (2015) studied the economic impact of different carbon tax revenue recycling schemes in China. Ricci (2012) studied providing adequate economic incentives for bioenergies with CO_2 capture and geological storage. Saveyn *et al.* (2012) performed economic analysis of a low carbon path to 2050 as a case for China, India and Japan.

Wittwer and Banerjee (2015) studied investing in irrigation development in North West Queensland, Australia. Li *et al.* (2015) explored economic impacts of total water use control in the Heihe River Basin in

Northwestern China. He *et al.* (2007) constructed a dynamic computable general equilibrium model and performed sensitivity analysis for shadow price of water resource in China.

Mai *et al.* (2014) studied the economic effects of facilitating the flow of rural workers to urban employment in China. Cockburn *et al.* (2014) studied impacts of the global economic crisis and national policy responses on children in Cameroon.

3. Framework, methodology, and results

The standard version of GDyn is a recursive-dynamic extension of the standard GTAP (Hertel, 1997), developed for better treatment of medium and long-term simulations, as it strengthens the investment side of the modeling framework to enable international capital mobility (Ianchovichina & Walmsley, 2012).

GDyn extends the standard, comparative static version of the GTAP model by introducing

- international capital mobility,
- endogenous capital accumulation and
- adaptive expectations theory of investment in a recursive dynamics setting.

The model determines the global markets for products, so that the balance is determined by the conditions of all countries' supply and demand. Demand for imports of a country is determined by its demand for imported inputs and goods consumed by final demand.

In this study, a fictitious trade and its 3-year-term potential effects on logistics services are simulated. The results indicate that some logistics services are much more sensitive to changes in some economies. The results indicate that various logistics services to/from economies exhibit increases or decreases (See Tables 1 through 9).

Table 1. Food – 2015 (percentage changes)

qxs[Food**]	1 Australia	2 NZ	3 China	4 Japan	5 Korea	6 SA	7 Canada	8 US	9 SAM	10 Austria	11 Denmark	12 Finland	13 France	14 UK	15 Ireland	16 Italy	17 Netherl	18 Portugal	19 Sweden	20 Eur	21 Turkey	22 ROW
1 Australia	1.8	1.8	4.3	3.9	2.7	4.2	0.7	0.7	0.9	1.9	1.9	2.1	1.3	1.2	2.1	1.0	1.4	2.4	1.8	1.8	1.8	2.6
2 NZ	2.3	2.3	4.7	4.3	3.1	4.6	1.1	1.1	1.4	2.4	2.3	2.5	1.7	1.6	2.5	1.4	1.8	2.8	2.2	2.2	2.3	3.0
3 China	4.2	4.2	6.8	6.3	4.9	6.5	3.0	3.0	3.3	4.5	4.4	4.6	3.7	3.6	4.6	3.4	3.8	4.9	4.3	4.2	4.2	5.0
4 Japan	-5.2	-5.2	-2.8	-3.6	-4.3	-2.8	-6.2	-6.2	-6.1	-5.4	-5.4	-5.2	-5.9	-5.9	-5.1	-6.2	-5.6	-4.8	-5.4	-5.4	-5.2	-4.5
5 Korea	-1.0	-1.0	1.6	1.1	-0.3	1.4	-2.1	-2.1	-1.9	-1.0	-1.0	-0.7	-1.6	-1.7	-0.7	-1.8	-1.4	-0.4	-1.1	-1.0	-1.1	-0.2
6 SA	-0.1	-0.1	2.4	2.0	0.9	2.3	-1.2	-1.2	-0.9	0.0	0.0	0.2	-0.6	-0.7	0.2	-0.9	-0.4	0.5	-0.1	-0.1	-0.1	0.7
7 Canada	1.9	1.9	4.3	3.9	2.7	4.2	0.7	0.7	1.0	2.0	1.9	2.2	1.3	1.2	2.1	1.0	1.5	2.5	1.9	1.8	1.9	2.6
8 US	3.4	3.5	6.0	5.5	4.3	5.8	2.4	2.5	2.6	3.6	3.5	3.8	2.9	2.8	3.6	2.6	3.1	4.0	3.5	3.4	3.4	4.2
9 SAM	4.1	4.1	6.5	6.2	4.9	6.4	2.9	2.9	3.1	4.3	4.0	4.4	3.4	3.4	4.2	3.1	3.6	4.6	4.0	4.0	4.0	4.8
10 Austria	1.4	1.4	3.9	3.5	2.3	3.8	0.3	0.3	0.5	1.5	1.5	1.7	0.9	0.8	1.7	0.6	1.0	2.0	1.4	1.4	1.4	2.2
11 Denmark	0.7	0.7	3.1	2.7	1.5	3.0	-0.5	-0.5	-0.2	0.7	0.7	0.9	0.1	0.0	0.9	-0.2	0.2	1.2	0.6	0.6	0.6	1.4
12 Finland	0.9	0.9	3.4	2.9	1.8	3.2	-0.2	-0.2	0.1	1.0	0.9	1.2	0.3	0.3	1.2	0.1	0.5	1.5	0.9	0.9	0.9	1.7
13 France	2.1	2.1	4.5	4.1	3.0	4.4	0.9	0.9	1.2	2.2	2.1	2.4	1.6	1.5	2.4	1.3	1.7	2.7	2.1	2.1	2.1	2.8
14 UK	1.6	1.6	4.1	3.6	2.4	3.9	0.4	0.4	0.7	1.7	1.6	1.9	1.0	1.0	1.9	0.8	1.2	2.2	1.6	1.5	1.6	2.3
15 Ireland	3.6	3.6	6.0	5.6	4.5	5.9	2.4	2.4	2.6	3.7	3.7	3.9	3.1	3.0	4.0	2.8	3.3	4.2	3.6	3.6	3.5	4.3
16 Italy	2.7	2.7	5.2	4.8	3.6	5.0	1.5	1.5	1.8	2.8	2.8	3.0	2.2	2.1	3.0	1.9	2.3	3.3	2.7	2.7	2.7	3.4
17 Netherl	0.6	0.6	3.1	2.7	1.5	2.9	-0.5	-0.5	-0.3	0.7	0.6	0.9	0.0	-0.1	0.9	-0.3	0.2	1.2	0.6	0.6	0.6	1.4
18 Portugal	-0.1	-0.1	2.4	2.0	0.8	2.2	-1.2	-1.2	-0.9	0.0	0.0	0.2	-0.6	-0.7	0.2	-0.9	-0.5	0.5	-0.1	-0.1	-0.1	0.7
19 Sweden	0.4	0.4	2.8	2.4	1.3	2.7	-0.8	-0.8	-0.5	0.5	0.4	0.7	-0.2	-0.3	0.6	-0.5	0.0	0.9	0.3	0.4	0.4	1.2
20 Eur	0.8	0.8	3.3	2.9	1.7	3.2	-0.3	-0.3	0.0	1.0	0.9	1.1	0.3	0.2	1.1	0.0	0.5	1.4	0.8	0.8	0.8	1.6
21 Turkey	3.1	3.0	5.7	5.2	4.0	5.5	2.0	2.0	2.2	3.2	3.2	3.4	2.5	2.4	3.4	2.3	2.7	3.7	3.1	3.0	3.2	3.8
22 ROW	1.6	1.6	4.1	3.7	2.5	4.0	0.5	0.5	0.7	1.7	1.7	1.9	1.1	1.0	1.9	0.8	1.2	2.2	1.6	1.6	1.6	2.4

Table 2. Mnfcs – 2015 (percentage changes)

qxs[Mnfcs**]	1 Australia	2 NZ	3 China	4 Japan	5 Korea	6 SA	7 Canada	8 US	9 SAM	10 Austria	11 Denmark	12 Finland	13 France	14 UK	15 Ireland	16 Italy	17 Netherl	18 Portugal	19 Sweden	20 Eur	21 Turkey	22 ROW
1 Australia	3.5	3.6	6.5	6.0	5.5	5.3	1.6	1.9	1.8	4.4	3.9	4.5	3.1	3.4	5.2	3.2	3.5	4.3	4.2	3.9	4.8	4.3
2 NZ	3.4	3.5	6.4	6.0	5.3	5.2	1.5	1.8	1.8	4.3	4.0	4.8	3.2	3.4	5.1	3.1	3.6	4.4	4.2	3.9	4.9	4.2
3 China	4.7	4.9	8.0	7.5	6.9	6.7	2.9	3.2	3.2	5.6	5.3	6.1	4.5	4.7	6.6	4.5	5.0	5.7	5.6	5.3	6.3	5.6
4 Japan	-7.1	-7.0	-4.1	-5.0	-5.2	-5.4	-8.8	-8.5	-8.6	-6.4	-6.6	-6.0	-7.3	-7.2	-5.7	-7.4	-7.0	-6.3	-6.4	-6.7	-5.8	-6.4
5 Korea	2.6	2.6	5.6	5.2	4.6	4.4	0.7	1.0	1.0	3.4	3.1	3.9	2.4	2.5	4.2	2.3	2.7	3.5	3.4	3.1	4.0	3.4
6 SA	5.5	5.6	8.6	8.2	7.5	7.4	3.7	4.0	3.9	6.4	6.1	6.9	5.3	5.5	7.3	5.2	5.7	6.5	6.4	6.1	7.0	6.4
7 Canada	2.9	3.0	6.0	5.5	5.0	4.8	1.1	1.4	1.3	3.8	3.5	4.3	2.7	2.9	4.6	2.7	3.1	3.8	3.7	3.4	4.4	3.8
8 US	5.0	5.1	8.1	7.7	7.1	6.9	3.2	3.6	3.4	5.9	5.6	6.4	4.8	5.0	6.7	4.7	5.2	5.9	5.9	5.5	6.5	5.9
9 SAM	7.8	7.8	10.9	10.3	9.8	9.7	6.0	6.3	6.1	8.8	8.0	9.1	7.5	7.8	9.5	7.4	7.9	8.6	8.6	8.3	9.2	8.7
10 Austria	2.1	2.2	5.1	4.7	4.1	3.9	0.2	0.6	0.5	2.9	2.6	3.4	1.9	2.0	3.7	1.8	2.2	3.0	2.9	2.6	3.5	2.9
11 Denmark	1.1	1.2	4.1	3.7	3.1	2.9	-0.7	-0.4	-0.5	2.0	1.6	2.4	0.9	1.1	2.7	0.8	1.3	2.0	1.9	1.6	2.6	1.9
12 Finland	1.4	1.5	4.4	4.0	3.4	3.2	-0.5	-0.1	-0.2	2.2	1.9	2.7	1.2	1.3	3.0	1.1	1.5	2.3	2.2	1.9	2.8	2.2
13 France	2.8	2.9	5.9	5.5	4.9	4.7	1.0	1.3	1.3	3.7	3.4	4.2	2.7	2.8	4.5	2.6	3.0	3.8	3.7	3.4	4.3	3.7
14 UK	2.2	2.3	5.2	4.8	4.2	4.0	0.4	0.7	0.6	3.1	2.7	3.5	2.0	2.2	3.9	1.9	2.3	3.1	3.0	2.7	3.6	3.0
15 Ireland	4.8	4.8	7.9	7.5	6.9	6.6	2.9	3.2	3.2	5.7	5.4	6.2	4.6	4.8	6.5	4.5	5.0	5.8	5.7	5.3	6.2	5.6
16 Italy	4.4	4.5	7.5	7.1	6.5	6.3	2.6	2.9	2.8	5.4	5.1	5.8	4.3	4.5	6.2	4.3	4.7	5.5	5.3	5.0	5.9	5.3
17 Netherl	1.2	1.3	4.3	3.9	3.3	3.1	-0.6	-0.3	-0.3	2.1	1.8	2.5	1.0	1.2	2.9	1.0	1.4	2.2	2.0	1.7	2.7	2.1
18 Portugal	1.1	1.2	4.1	3.7	3.1	2.9	-0.7	-0.4	-0.5	1.9	1.6	2.4	0.9	1.0	2.7	0.8	1.2	2.0	1.9	1.6	2.5	1.9
19 Sweden	0.9	1.0	3.9	3.5	2.9	2.7	-0.9	-0.6	-0.6	1.8	1.5	2.2	0.7	0.9	2.6	0.7	1.1	1.8	1.7	1.4	2.4	1.7
20 Eur	1.1	1.2	4.1	3.7	3.1	2.9	-0.7	-0.4	-0.4	2.0	1.6	2.4	0.9	1.1	2.7	0.8	1.3	2.0	1.9	1.6	2.6	1.9
21 Turkey	7.0	7.2	10.0	9.8	9.1	8.8	5.1	5.4	5.4	7.9	7.6	8.4	6.8	7.0	8.7	6.7	7.2	8.0	7.9	7.6	8.8	7.9
22 ROW	2.5	2.6	5.6	5.2	4.6	4.4	0.7	1.0	1.0	3.4	3.1	3.9	2.3	2.5	4.2	2.3	2.7	3.5	3.4	3.1	4.0	3.4

Table 3. Svces – 2015 (percentage changes)

qxs[Svces**]	1 Australia	2 NZ	3 China	4 Japan	5 Korea	6 SA	7 Canada	8 US	9 SAM	10 Austria	11 Denmark	12 Finland	13 France	14 UK	15 Ireland	16 Italy	17 Netherl	18 Portugal	19 Sweden	20 Eur	21 Turkey	22 ROW
1 Australia	3.2	3.2	6.7	7.1	5.7	4.2	2.8	2.3	2.2	3.4	3.7	4.5	2.5	3.3	4.2	2.4	4.0	4.3	4.3	3.9	3.7	3.6
2 NZ	3.0	2.9	6.4	6.8	5.4	3.9	2.5	2.0	1.9	3.2	3.4	4.2	2.2	3.0	3.9	2.1	3.7	4.0	4.0	3.6	3.4	3.3
3 China	4.2	4.2	7.7	8.1	6.7	5.2	3.8	3.2	3.2	4.4	4.6	5.5	3.4	4.3	5.2	3.4	5.0	5.3	5.3	4.9	4.7	4.6
4 Japan	-5.1	-5.1	-1.9	-1.5	-2.8	-4.2	-5.4	-5.9	-6.0	-4.9	-4.7	-3.9	-5.8	-5.0	-4.2	-5.8	-4.3	-4.1	-4.0	-4.4	-4.6	-4.8
5 Korea	2.2	2.2	5.7	6.1	4.7	3.2	1.8	1.3	1.2	2.4	2.6	3.5	1.5	2.3	3.2	1.4	3.0	3.3	3.3	2.9	2.7	2.6
6 SA	5.4	5.3	8.9	9.3	7.9	6.4	4.9	4.4	4.3	5.6	5.8	6.6	4.6	5.4	6.3	4.5	6.2	6.5	6.5	6.1	5.8	5.7
7 Canada	2.4	2.4	5.8	6.2	4.8	3.3	2.0	1.4	1.4	2.6	2.8	3.6	1.6	2.4	3.3	1.5	3.2	3.4	3.5	3.0	2.8	2.7
8 US	4.2	4.2	7.7	8.1	6.7	5.2	3.8	3.2	3.2	4.4	4.6	5.5	3.4	4.3	5.2	3.4	5.0	5.3	5.3	4.9	4.7	4.5
9 SAM	6.3	6.3	9.9	10.3	8.9	7.3	5.9	5.3	5.3	6.5	6.8	7.6	5.5	6.4	7.3	5.5	7.1	7.4	7.5	7.0	6.8	6.7
10 Austria	2.1	2.1	5.5	6.0	4.6	3.1	1.7	1.2	1.1	2.3	2.5	3.3	1.3	2.2	3.0	1.3	2.9	3.2	3.2	2.8	2.5	2.4
11 Denmark	1.0	1.0	4.4	4.8	3.4	1.9	0.6	0.1	0.0	1.2	1.4	2.2	0.2	1.0	1.9	0.2	1.8	2.1	2.1	1.7	1.4	1.3
12 Finland	1.3	1.3	4.7	5.1	3.7	2.3	0.9	0.4	0.3	1.5	1.7	2.5	0.6	1.4	2.2	0.5	2.1	2.4	2.4	2.0	1.8	1.6
13 France	3.3	3.3	6.8	7.2	5.8	4.3	2.9	2.4	2.3	3.5	3.8	4.6	2.6	3.4	4.3	2.5	4.1	4.4	4.4	4.0	3.8	3.7
14 UK	2.0	2.0	5.4	5.8	4.4	3.0	1.6	1.0	1.0	2.2	2.4	3.2	1.2	2.0	2.9	1.2	2.8	3.1	3.1	2.7	2.4	2.3
15 Ireland	4.4	4.4	7.9	8.3	6.9	5.4	4.0	3.4	3.4	4.6	4.8	5.7	3.6	4.4	5.3	3.6	5.2	5.5	5.5	5.1	4.8	4.7
16 Italy	4.5	4.5	8.1	8.5	7.1	5.5	4.1	3.6	3.5	4.7	5.0	5.8	3.8	4.6	5.5	3.7	5.3	5.6	5.7	5.2	5.0	4.9
17 Netherl	0.9	0.9	4.3	4.7	3.3	1.9	0.5	0.0	-0.1	1.1	1.3	2.1	0.2	1.0	1.8	0.1	1.7	2.0	2.0	1.6	1.4	1.2
18 Portugal	1.1	1.1	4.5	4.9	3.5	2.0	0.7	0.1	0.1	1.3	1.5	2.3	0.3	1.1	2.0	0.3	1.9	2.1	2.1	1.7	1.5	1.4
19 Sweden	0.9	0.9	4.3	4.7	3.3	1.8	0.5	-0.1	-0.1	1.1	1.3	2.1	0.1	0.9	1.8	0.1	1.6	1.9	1.9	1.5	1.3	1.2
20 Eur	1.0	1.0	4.4	4.8	3.4	1.9	0.6	0.1	0.0	1.2	1.4	2.2	0.2	1.0	1.9	0.2	1.8	2.0	2.1	1.7	1.4	1.3
21 Turkey	6.4	6.3	9.9	10.4	8.9	7.4	5.9	5.4	5.3	6.6	6.8	7.7	5.6	6.4	7.3	5.5	7.2	7.5	7.5	7.1	6.8	6.7
22 ROW	3.0	3.0	6.5	6.9	5.5	4.0	2.6	2.1	2.0	3.2	3.4	4.3	2.3	3.1	4.0	2.2	3.8	4.1	4.1	3.7	3.5	3.4

Table 4. Food – 2016 (percentage changes)

qxs[Food**]	1 Australia	2 NZ	3 China	4 Japan	5 Korea	6 SA	7 Canada	8 US	9 SAM	10 Austria	11 Denmark	12 Finland	13 France	14 UK	15 Ireland	16 Italy	17 Netherl	18 Portugal	19 Sweden	20 Eur	21 Turkey	22 ROW
1 Australia	1.8	1.8	4.3	3.8	2.7	4.0	0.7	0.8	1.1	1.8	1.8	2.0	1.2	1.2	2.1	0.9	1.4	2.3	1.7	1.7	1.9	2.5
2 NZ	2.3	2.3	4.8	4.2	3.2	4.4	1.1	1.2	1.5	2.3	2.2	2.4	1.6	1.6	2.4	1.4	1.8	2.7	2.1	2.1	2.3	3.0
3 China	4.2	4.2	6.8	6.2	5.0	6.4	3.0	3.1	3.4	4.4	4.3	4.6	3.6	3.6	4.6	3.4	3.8	4.8	4.2	4.1	4.2	4.9
4 Japan	-5.0	-5.1	-2.6	-3.5	-4.1	-2.8	-6.0	-5.9	-5.9	-5.3	-5.3	-5.1	-5.8	-5.8	-4.9	-6.0	-5.5	-4.7	-5.3	-5.3	-5.0	-4.4
5 Korea	-1.0	-1.0	1.6	1.0	-0.2	1.2	-2.0	-2.0	-1.8	-1.1	-1.1	-0.8	-1.7	-1.7	-0.7	-1.9	-1.4	-0.5	-1.2	-1.1	-1.0	-0.3
6 SA	0.3	0.3	2.8	2.3	1.3	2.5	-0.7	-0.7	-0.4	0.3	0.3	0.5	-0.3	-0.3	0.6	-0.6	-0.1	0.8	0.2	0.2	0.4	1.0
7 Canada	1.8	1.8	4.3	3.8	2.7	4.0	0.8	0.8	1.1	1.8	1.8	2.0	1.2	1.2	2.1	0.9	1.4	2.3	1.7	1.7	1.9	2.5
8 US	3.2	3.3	5.8	5.2	4.1	5.4	2.3	2.3	2.5	3.3	3.2	3.5	2.6	2.5	3.4	2.4	2.8	3.7	3.2	3.1	3.3	4.0
9 SAM	4.0	4.0	6.5	6.0	4.8	6.1	2.9	2.9	3.2	4.1	3.8	4.2	3.2	3.2	4.1	3.0	3.4	4.3	3.8	3.8	3.9	4.7
10 Austria	1.5	1.5	4.0	3.5	2.4	3.7	0.4	0.5	0.7	1.5	1.5	1.7	0.9	0.8	1.8	0.6	1.1	2.0	1.4	1.4	1.5	2.2
11 Denmark	0.7	0.8	3.3	2.7	1.7	2.9	-0.3	-0.3	0.0	0.8	0.7	1.0	0.1	0.1	1.0	-0.1	0.3	1.2	0.6	0.6	0.8	1.5
12 Finland	1.0	1.0	3.5	3.0	1.9	3.2	0.0	0.0	0.3	1.0	1.0	1.2	0.4	0.4	1.3	0.2	0.6	1.5	0.9	0.9	1.1	1.8
13 France	2.1	2.2	4.7	4.1	3.1	4.3	1.1	1.1	1.4	2.2	2.1	2.4	1.6	1.5	2.4	1.3	1.7	2.7	2.1	2.1	2.2	2.9
14 UK	1.6	1.6	4.1	3.6	2.6	3.8	0.6	0.6	0.9	1.6	1.6	1.8	1.0	1.0	1.9	0.8	1.2	2.1	1.5	1.5	1.7	2.4
15 Ireland	3.7	3.7	6.1	5.6	4.5	5.8	2.5	2.6	2.8	3.7	3.7	3.9	3.0	3.0	4.0	2.8	3.3	4.1	3.6	3.5	3.7	4.3
16 Italy	2.8	2.8	5.3	4.8	3.7	5.0	1.7	1.7	2.0	2.8	2.8	3.0	2.2	2.2	3.1	2.0	2.4	3.3	2.7	2.7	2.8	3.5
17 Netherl	0.7	0.7	3.2	2.7	1.6	2.9	-0.4	-0.3	-0.1	0.7	0.6	0.9	0.0	0.0	0.9	-0.2	0.2	1.2	0.6	0.6	0.7	1.4
18 Portugal	0.1	0.1	2.6	2.0	1.0	2.3	-0.9	-0.9	-0.6	0.1	0.1	0.3	-0.5	-0.6	0.4	-0.8	-0.3	0.6	0.0	0.0	0.2	0.9
19 Sweden	0.5	0.5	3.0	2.5	1.4	2.7	-0.5	-0.5	-0.2	0.5	0.5	0.8	-0.1	-0.2	0.8	-0.4	0.1	1.0	0.4	0.4	0.6	1.3
20 Eur	1.0	1.0	3.5	2.9	1.9	3.2	-0.1	-0.1	0.2	1.0	0.9	1.2	0.4	0.3	1.2	0.1	0.5	1.5	0.9	0.9	1.0	1.7
21 Turkey	3.2	3.1	5.8	5.2	4.2	5.4	2.1	2.1	2.4	3.1	3.2	3.4	2.5	2.5	3.4	2.3	2.7	3.7	3.0	3.0	3.3	3.9
22 ROW	1.7	1.7	4.2	3.7	2.6	3.9	0.6	0.7	0.9	1.7	1.7	1.9	1.1	1.0	2.0	0.8	1.3	2.2	1.6	1.6	1.7	2.4

Table 5. Mnfcs – 2016 (percentage changes)

qxs[Mnfcs**]	1 Australia	2 NZ	3 China	4 Japan	5 Korea	6 SA	7 Canada	8 US	9 SAM	10 Austria	11 Denmark	12 Finland	13 France	14 UK	15 Ireland	16 Italy	17 Netherl	18 Portugal	19 Sweden	20 Eur	21 Turkey	22 ROW
1 Australia	3.4	3.5	6.4	5.9	5.3	5.1	1.7	1.9	1.8	4.2	3.7	4.3	3.0	3.3	5.0	3.0	3.4	4.2	4.1	3.8	4.7	4.1
2 NZ	3.4	3.5	6.4	5.9	5.2	5.1	1.6	1.9	1.9	4.1	3.8	4.6	3.1	3.3	5.0	3.0	3.5	4.2	4.1	3.8	4.7	4.1
3 China	4.7	4.8	7.9	7.4	6.8	6.6	3.0	3.2	3.2	5.4	5.1	5.9	4.4	4.6	6.5	4.3	4.8	5.5	5.4	5.1	6.1	5.5
4 Japan	-6.9	-6.9	-3.9	-4.8	-5.1	-5.3	-8.5	-8.3	-8.3	-6.3	-6.5	-5.9	-7.2	-7.0	-5.5	-7.3	-6.9	-6.2	-6.3	-6.6	-5.7	-6.3
5 Korea	2.6	2.7	5.6	5.2	4.6	4.4	0.9	1.2	1.1	3.4	3.1	3.8	2.3	2.5	4.2	2.3	2.7	3.4	3.3	3.0	4.0	3.3
6 SA	5.4	5.5	8.5	8.0	7.3	7.2	3.7	4.0	3.9	6.2	5.9	6.7	5.1	5.4	7.2	5.0	5.5	6.3	6.2	5.8	6.8	6.2
7 Canada	2.8	2.9	5.9	5.4	4.8	4.6	1.2	1.4	1.4	3.6	3.3	4.1	2.6	2.8	4.5	2.5	2.9	3.7	3.6	3.3	4.2	3.6
8 US	4.7	4.8	7.8	7.4	6.7	6.5	3.0	3.4	3.2	5.5	5.2	6.0	4.4	4.7	6.4	4.4	4.8	5.5	5.5	5.1	6.1	5.5
9 SAM	7.5	7.6	10.7	10.1	9.5	9.4	5.9	6.2	6.0	8.5	7.7	8.8	7.2	7.5	9.2	7.1	7.6	8.3	8.3	7.9	8.9	8.3
10 Austria	2.1	2.2	5.2	4.8	4.1	3.9	0.5	0.7	0.7	2.9	2.6	3.4	1.9	2.1	3.8	1.8	2.2	3.0	2.9	2.6	3.5	2.9
11 Denmark	1.2	1.3	4.2	3.8	3.1	2.9	-0.5	-0.2	-0.3	1.9	1.6	2.4	0.9	1.1	2.8	0.8	1.3	2.0	1.9	1.6	2.6	1.9
12 Finland	1.5	1.6	4.6	4.2	3.5	3.3	-0.1	0.1	0.1	2.3	2.0	2.7	1.2	1.4	3.1	1.2	1.6	2.4	2.3	1.9	2.9	2.3
13 France	2.9	3.0	6.0	5.6	4.9	4.7	1.2	1.5	1.5	3.7	3.4	4.2	2.7	2.9	4.6	2.6	3.0	3.8	3.7	3.4	4.3	3.7
14 UK	2.2	2.3	5.3	4.8	4.2	4.0	0.5	0.8	0.8	3.0	2.7	3.5	2.0	2.2	3.9	1.9	2.3	3.1	3.0	2.6	3.6	3.0
15 Ireland	4.8	4.8	7.9	7.5	6.8	6.6	3.0	3.3	3.3	5.6	5.3	6.1	4.5	4.7	6.5	4.5	4.9	5.7	5.6	5.2	6.2	5.5
16 Italy	4.4	4.5	7.5	7.1	6.5	6.3	2.7	3.0	3.0	5.3	5.0	5.8	4.2	4.4	6.2	4.2	4.6	5.4	5.3	4.9	5.9	5.2
17 Netherl	1.3	1.4	4.4	3.9	3.3	3.1	-0.3	-0.1	-0.1	2.1	1.8	2.5	1.1	1.3	2.9	1.0	1.4	2.2	2.1	1.7	2.7	2.1
18 Portugal	1.2	1.2	4.2	3.8	3.1	2.9	-0.5	-0.2	-0.3	1.9	1.6	2.4	0.9	1.1	2.8	0.8	1.2	2.0	1.9	1.6	2.6	1.9
19 Sweden	1.0	1.1	4.0	3.6	3.0	2.8	-0.6	-0.4	-0.4	1.8	1.5	2.2	0.8	0.9	2.6	0.7	1.1	1.8	1.7	1.4	2.4	1.8
20 Eur	1.2	1.3	4.2	3.8	3.2	3.0	-0.5	-0.2	-0.2	2.0	1.7	2.4	0.9	1.1	2.8	0.9	1.3	2.0	1.9	1.6	2.6	1.9
21 Turkey	6.8	7.0	9.9	9.6	8.9	8.6	5.1	5.4	5.4	7.7	7.4	8.1	6.6	6.8	8.5	6.5	7.0	7.7	7.7	7.3	8.6	7.7
22 ROW	2.7	2.7	5.7	5.3	4.6	4.4	1.0	1.2	1.2	3.4	3.1	3.9	2.4	2.6	4.3	2.3	2.8	3.5	3.4	3.1	4.0	3.4

Table 6. Svces – 2016 (percentage changes)

qxs[Svces**]	1 Australia	2 NZ	3 China	4 Japan	5 Korea	6 SA	7 Canada	8 US	9 SAM	10 Austria	11 Denmark	12 Finland	13 France	14 UK	15 Ireland	16 Italy	17 Netherl	18 Portugal	19 Sweden	20 Eur	21 Turkey	22 ROW
1 Australia	3.2	3.2	6.7	7.0	5.6	4.2	2.8	2.3	2.2	3.3	3.6	4.4	2.4	3.2	4.1	2.3	3.9	4.2	4.2	3.8	3.7	3.5
2 NZ	3.0	2.9	6.4	6.7	5.4	4.0	2.5	2.1	2.0	3.1	3.3	4.1	2.2	3.0	3.9	2.1	3.7	4.0	3.9	3.5	3.4	3.2
3 China	4.2	4.1	7.7	8.0	6.6	5.2	3.8	3.3	3.2	4.3	4.5	5.3	3.4	4.2	5.1	3.3	4.9	5.2	5.2	4.8	4.6	4.5
4 Japan	-4.9	-4.9	-1.7	-1.4	-2.7	-4.0	-5.3	-5.7	-5.8	-4.8	-4.6	-3.8	-5.6	-4.9	-4.1	-5.7	-4.2	-4.0	-4.0	-4.4	-4.5	-4.6
5 Korea	2.3	2.3	5.8	6.0	4.7	3.3	1.9	1.4	1.3	2.4	2.6	3.4	1.5	2.3	3.2	1.4	3.0	3.3	3.3	2.9	2.8	2.6
6 SA	5.3	5.2	8.8	9.1	7.7	6.3	4.8	4.3	4.3	5.4	5.6	6.4	4.4	5.3	6.2	4.4	6.0	6.3	6.3	5.8	5.7	5.5
7 Canada	2.4	2.3	5.8	6.1	4.8	3.4	2.0	1.5	1.4	2.5	2.7	3.5	1.6	2.4	3.3	1.5	3.1	3.4	3.4	2.9	2.8	2.6
8 US	4.0	4.0	7.5	7.8	6.5	5.0	3.6	3.1	3.0	4.2	4.4	5.2	3.2	4.0	4.9	3.1	4.7	5.0	5.0	4.6	4.5	4.3
9 SAM	6.2	6.2	9.8	10.1	8.7	7.3	5.8	5.3	5.2	6.3	6.6	7.4	5.4	6.2	7.1	5.3	6.9	7.2	7.2	6.8	6.7	6.5
10 Austria	2.2	2.2	5.7	5.9	4.6	3.2	1.8	1.3	1.3	2.3	2.6	3.3	1.4	2.2	3.1	1.4	2.9	3.2	3.2	2.8	2.7	2.5
11 Denmark	1.1	1.1	4.5	4.8	3.5	2.1	0.7	0.2	0.2	1.2	1.4	2.2	0.3	1.1	2.0	0.3	1.8	2.1	2.1	1.7	1.6	1.4
12 Finland	1.5	1.4	4.9	5.2	3.8	2.5	1.0	0.6	0.5	1.6	1.8	2.6	0.7	1.5	2.3	0.6	2.2	2.4	2.4	2.0	1.9	1.7
13 France	3.4	3.4	6.9	7.2	5.8	4.4	3.0	2.5	2.4	3.5	3.7	4.5	2.6	3.4	4.3	2.5	4.1	4.4	4.4	4.0	3.9	3.7
14 UK	2.1	2.0	5.5	5.8	4.5	3.1	1.7	1.2	1.1	2.2	2.4	3.2	1.3	2.1	3.0	1.2	2.8	3.1	3.1	2.6	2.5	2.3
15 Ireland	4.5	4.4	8.0	8.3	6.9	5.5	4.0	3.6	3.5	4.6	4.8	5.6	3.7	4.5	5.4	3.6	5.2	5.5	5.5	5.1	4.9	4.7
16 Italy	4.6	4.5	8.1	8.4	7.0	5.6	4.1	3.7	3.6	4.7	4.9	5.7	3.8	4.6	5.5	3.7	5.3	5.6	5.6	5.1	5.0	4.8
17 Netherl	1.0	1.0	4.5	4.7	3.4	2.0	0.6	0.2	0.1	1.2	1.4	2.2	0.3	1.0	1.9	0.2	1.7	2.0	2.0	1.6	1.5	1.3
18 Portugal	1.2	1.2	4.6	4.9	3.6	2.2	0.8	0.3	0.2	1.3	1.5	2.3	0.4	1.2	2.1	0.3	1.9	2.2	2.2	1.8	1.6	1.5
19 Sweden	1.0	1.0	4.4	4.7	3.4	2.0	0.6	0.1	0.1	1.1	1.3	2.1	0.2	1.0	1.9	0.2	1.7	2.0	2.0	1.6	1.5	1.3
20 Eur	1.2	1.1	4.6	4.8	3.5	2.2	0.7	0.3	0.2	1.3	1.5	2.3	0.4	1.2	2.0	0.3	1.9	2.1	2.1	1.7	1.6	1.4
21 Turkey	6.3	6.2	9.9	10.1	8.8	7.3	5.8	5.4	5.3	6.4	6.6	7.5	5.5	6.3	7.2	5.4	7.0	7.3	7.3	6.9	6.8	6.6
22 ROW	3.2	3.1	6.6	6.9	5.6	4.2	2.7	2.3	2.2	3.3	3.5	4.3	2.4	3.2	4.1	2.3	3.9	4.1	4.1	3.7	3.6	3.4

Table 7. Food – 2017 (percentage changes)

qxs[Food**]	1 Australia	2 NZ	3 China	4 Japan	5 Korea	6 SA	7 Canada	8 US	9 SAM	10 Austria	11 Denmark	12 Finland	13 France	14 UK	15 Ireland	16 Italy	17 Netherl	18 Portugal	19 Sweden	20 Eur	21 Turkey	22 ROW
1 Australia	1.8	1.8	4.3	3.7	2.8	3.9	0.8	0.8	1.2	1.7	1.7	2.0	1.1	1.1	2.0	0.9	1.3	2.2	1.6	1.6	1.9	2.5
2 NZ	2.3	2.3	4.8	4.1	3.2	4.3	1.2	1.3	1.6	2.2	2.2	2.4	1.5	1.5	2.4	1.3	1.7	2.6	2.0	2.0	2.4	2.9
3 China	4.2	4.2	6.8	6.1	5.0	6.2	3.1	3.2	3.5	4.3	4.2	4.5	3.5	3.5	4.5	3.3	3.7	4.7	4.1	4.0	4.2	4.9
4 Japan	-4.9	-4.9	-2.4	-3.4	-3.9	-2.7	-5.7	-5.7	-5.6	-5.2	-5.1	-4.9	-5.7	-5.6	-4.8	-5.9	-5.4	-4.7	-5.3	-5.2	-4.8	-4.3
5 Korea	-1.0	-1.0	1.6	0.9	-0.2	1.1	-1.9	-1.9	-1.7	-1.2	-1.1	-0.9	-1.7	-1.8	-0.8	-1.9	-1.5	-0.7	-1.3	-1.2	-1.0	-0.3
6 SA	0.6	0.7	3.2	2.5	1.6	2.7	-0.3	-0.3	0.0	0.5	0.6	0.8	0.0	-0.1	0.9	-0.3	0.2	1.1	0.5	0.5	0.7	1.4
7 Canada	1.8	1.8	4.3	3.7	2.7	3.8	0.8	0.8	1.1	1.7	1.7	1.9	1.1	1.1	2.0	0.9	1.3	2.2	1.6	1.6	1.9	2.5
8 US	3.0	3.1	5.6	4.9	4.0	5.1	2.1	2.2	2.4	3.0	2.9	3.2	2.3	2.3	3.2	2.1	2.6	3.4	2.9	2.9	3.1	3.7
9 SAM	3.9	3.9	6.4	5.8	4.7	5.9	2.8	2.9	3.2	3.9	3.7	4.0	3.1	3.1	4.0	2.8	3.3	4.1	3.6	3.6	3.9	4.6
10 Austria	1.6	1.6	4.1	3.5	2.5	3.6	0.6	0.6	0.9	1.5	1.5	1.7	0.9	0.9	1.8	0.7	1.1	2.0	1.4	1.4	1.7	2.3
11 Denmark	0.8	0.9	3.4	2.7	1.8	2.9	-0.1	-0.1	0.2	0.8	0.7	1.0	0.2	0.1	1.1	-0.1	0.4	1.2	0.6	0.6	0.9	1.6
12 Finland	1.1	1.2	3.7	3.0	2.1	3.2	0.2	0.2	0.6	1.1	1.1	1.3	0.5	0.4	1.4	0.2	0.7	1.6	1.0	1.0	1.2	1.9
13 France	2.2	2.2	4.7	4.1	3.2	4.3	1.2	1.3	1.6	2.1	2.1	2.4	1.6	1.5	2.5	1.3	1.7	2.6	2.0	2.0	2.3	2.9
14 UK	1.7	1.7	4.2	3.6	2.6	3.7	0.7	0.7	1.1	1.6	1.6	1.8	1.0	1.0	1.9	0.8	1.2	2.1	1.5	1.5	1.8	2.4
15 Ireland	3.7	3.7	6.2	5.6	4.7	5.8	2.7	2.7	3.0	3.6	3.7	3.9	3.0	3.0	4.0	2.8	3.3	4.1	3.5	3.5	3.8	4.4
16 Italy	2.8	2.9	5.4	4.8	3.8	4.9	1.9	1.9	2.2	2.8	2.8	3.0	2.2	2.2	3.1	2.0	2.4	3.3	2.7	2.7	2.9	3.6
17 Netherl	0.8	0.8	3.3	2.7	1.7	2.8	-0.2	-0.1	0.2	0.7	0.7	0.9	0.1	0.1	1.0	-0.2	0.3	1.2	0.6	0.6	0.9	1.5
18 Portugal	0.3	0.3	2.8	2.1	1.2	2.3	-0.7	-0.6	-0.3	0.2	0.2	0.4	-0.4	-0.4	0.5	-0.6	-0.2	0.6	0.1	0.1	0.4	1.0
19 Sweden	0.7	0.7	3.2	2.6	1.6	2.7	-0.3	-0.3	0.1	0.6	0.6	0.8	0.0	-0.1	0.9	-0.3	0.2	1.0	0.4	0.5	0.8	1.4
20 Eur	1.1	1.1	3.6	3.0	2.1	3.2	0.1	0.2	0.5	1.0	1.0	1.3	0.4	0.4	1.3	0.2	0.6	1.5	0.9	0.9	1.2	1.8
21 Turkey	3.2	3.2	5.9	5.2	4.3	5.3	2.2	2.3	2.6	3.1	3.2	3.4	2.5	2.5	3.5	2.3	2.7	3.6	3.0	3.0	3.4	3.9
22 ROW	1.8	1.8	4.3	3.7	2.7	3.8	0.8	0.8	1.1	1.7	1.7	1.9	1.1	1.1	2.0	0.9	1.3	2.2	1.6	1.6	1.9	2.5

Table 8. Mnfcs – 2017 (percentage changes)

qxs[Mnfcs**]	1 Australia	2 NZ	3 China	4 Japan	5 Korea	6 SA	7 Canada	8 US	9 SAM	10 Austria	11 Denmark	12 Finland	13 France	14 UK	15 Ireland	16 Italy	17 Netherl	18 Portugal	19 Sweden	20 Eur	21 Turkey	22 ROW
1 Australia	3.4	3.4	6.3	5.8	5.2	5.0	1.7	1.9	1.9	4.0	3.6	4.2	2.9	3.1	4.9	2.9	3.3	4.0	3.9	3.6	4.5	4.0
2 NZ	3.3	3.4	6.3	5.8	5.1	5.0	1.8	1.9	1.9	4.0	3.7	4.5	3.0	3.2	4.9	2.9	3.4	4.1	4.0	3.6	4.6	4.0
3 China	4.6	4.7	7.8	7.3	6.7	6.5	3.0	3.3	3.3	5.3	4.9	5.8	4.3	4.5	6.4	4.2	4.7	5.4	5.3	5.0	6.0	5.3
4 Japan	-6.7	-6.7	-3.7	-4.6	-4.9	-5.1	-8.2	-8.0	-8.0	-6.2	-6.4	-5.7	-7.1	-6.9	-5.4	-7.1	-6.7	-6.1	-6.2	-6.5	-5.5	-6.1
5 Korea	2.7	2.7	5.7	5.2	4.6	4.4	1.1	1.3	1.3	3.3	3.0	3.8	2.3	2.5	4.2	2.3	2.7	3.4	3.3	3.0	4.0	3.3
6 SA	5.3	5.4	8.3	7.9	7.2	7.1	3.7	3.9	3.9	6.0	5.7	6.5	5.0	5.2	7.0	4.9	5.4	6.0	6.0	5.6	6.6	6.0
7 Canada	2.8	2.8	5.8	5.3	4.7	4.5	1.2	1.4	1.4	3.4	3.1	3.9	2.4	2.6	4.4	2.4	2.8	3.5	3.4	3.1	4.1	3.4
8 US	4.5	4.5	7.5	7.1	6.4	6.2	2.9	3.2	3.1	5.1	4.9	5.6	4.1	4.3	6.1	4.0	4.5	5.2	5.1	4.8	5.8	5.1
9 SAM	7.3	7.3	10.4	9.8	9.2	9.1	5.8	6.0	5.9	8.1	7.4	8.5	6.9	7.2	8.9	6.8	7.2	7.9	8.0	7.6	8.5	8.0
10 Austria	2.2	2.3	5.2	4.8	4.1	3.9	0.7	0.9	0.9	2.9	2.6	3.3	1.9	2.1	3.8	1.8	2.2	2.9	2.9	2.5	3.5	2.9
11 Denmark	1.3	1.3	4.2	3.8	3.1	2.9	-0.3	-0.1	-0.1	1.9	1.6	2.4	0.9	1.1	2.8	0.8	1.3	2.0	1.9	1.6	2.6	1.9
12 Finland	1.7	1.8	4.7	4.3	3.6	3.4	0.2	0.4	0.4	2.3	2.1	2.8	1.3	1.6	3.2	1.3	1.7	2.4	2.3	2.0	3.0	2.3
13 France	3.0	3.1	6.0	5.6	4.9	4.7	1.4	1.6	1.6	3.7	3.4	4.1	2.7	2.9	4.6	2.6	3.0	3.7	3.7	3.3	4.3	3.7
14 UK	2.3	2.3	5.3	4.8	4.2	4.0	0.7	0.9	0.9	2.9	2.6	3.4	1.9	2.1	3.9	1.9	2.3	3.0	2.9	2.6	3.6	2.9
15 Ireland	4.8	4.8	7.8	7.4	6.8	6.5	3.2	3.4	3.4	5.5	5.2	6.0	4.4	4.7	6.4	4.4	4.8	5.5	5.5	5.1	6.1	5.4
16 Italy	4.4	4.5	7.5	7.1	6.4	6.2	2.9	3.1	3.1	5.2	4.9	5.7	4.2	4.4	6.1	4.1	4.5	5.3	5.2	4.8	5.8	5.1
17 Netherl	1.4	1.5	4.4	4.0	3.3	3.1	-0.1	0.1	0.1	2.1	1.8	2.5	1.1	1.3	3.0	1.0	1.4	2.1	2.1	1.8	2.7	2.1
18 Portugal	1.3	1.4	4.3	3.8	3.2	3.0	-0.3	-0.1	-0.1	1.9	1.6	2.4	0.9	1.1	2.8	0.9	1.3	2.0	1.9	1.6	2.6	1.9
19 Sweden	1.2	1.2	4.1	3.7	3.1	2.9	-0.4	-0.2	-0.2	1.8	1.5	2.3	0.8	1.0	2.7	0.8	1.2	1.9	1.8	1.5	2.5	1.8
20 Eur	1.3	1.4	4.3	3.9	3.2	3.0	-0.2	0.0	0.0	2.0	1.7	2.5	1.0	1.2	2.9	0.9	1.4	2.0	2.0	1.7	2.6	2.0
21 Turkey	6.7	6.9	9.7	9.4	8.6	8.4	5.1	5.3	5.4	7.4	7.2	7.9	6.4	6.6	8.3	6.3	6.8	7.5	7.5	7.1	8.4	7.5
22 ROW	2.8	2.8	5.8	5.3	4.7	4.5	1.2	1.4	1.4	3.4	3.1	3.9	2.4	2.6	4.3	2.4	2.8	3.5	3.4	3.1	4.1	3.4

Table 9. Svces – 2017 (percentage changes)

qxs[Svces**]	1 Australia	2 NZ	3 China	4 Japan	5 Korea	6 SA	7 Canada	8 US	9 SAM	10 Austria	11 Denmark	12 Finland	13 France	14 UK	15 Ireland	16 Italy	17 Netherl	18 Portugal	19 Sweden	20 Eur	21 Turkey	22 ROW
1 Australia	3.2	3.1	6.7	6.8	5.5	4.2	2.8	2.3	2.3	3.2	3.5	4.2	2.4	3.1	4.1	2.3	3.8	4.1	4.1	3.7	3.7	3.4
2 NZ	3.0	2.9	6.5	6.6	5.3	4.0	2.5	2.1	2.1	3.0	3.2	4.0	2.1	2.9	3.8	2.1	3.6	3.9	3.9	3.4	3.5	3.2
3 China	4.1	4.1	7.7	7.8	6.5	5.2	3.7	3.3	3.2	4.2	4.4	5.2	3.3	4.1	5.0	3.3	4.8	5.1	5.0	4.6	4.6	4.4
4 Japan	-4.7	-4.8	-1.5	-1.4	-2.5	-3.7	-5.1	-5.5	-5.6	-4.7	-4.5	-3.8	-5.5	-4.8	-3.9	-5.5	-4.1	-3.9	-3.9	-4.3	-4.3	-4.5
5 Korea	2.4	2.3	5.8	6.0	4.7	3.4	1.9	1.5	1.5	2.4	2.6	3.4	1.5	2.3	3.2	1.5	3.0	3.3	3.2	2.8	2.8	2.6
6 SA	5.1	5.1	8.7	8.8	7.5	6.2	4.7	4.3	4.2	5.2	5.4	6.2	4.3	5.1	6.0	4.2	5.8	6.1	6.1	5.6	5.6	5.4
7 Canada	2.4	2.3	5.8	6.0	4.7	3.4	1.9	1.5	1.5	2.4	2.6	3.4	1.6	2.3	3.2	1.5	3.0	3.3	3.3	2.8	2.9	2.6
8 US	3.8	3.8	7.4	7.5	6.2	4.9	3.4	3.0	2.9	3.9	4.1	4.9	3.0	3.8	4.7	3.0	4.5	4.8	4.7	4.3	4.3	4.1
9 SAM	6.1	6.0	9.7	9.8	8.5	7.2	5.6	5.2	5.2	6.2	6.4	7.2	5.2	6.0	7.0	5.2	6.7	7.0	7.0	6.6	6.6	6.3
10 Austria	2.3	2.3	5.8	5.9	4.7	3.4	1.9	1.5	1.4	2.4	2.6	3.4	1.5	2.3	3.2	1.5	3.0	3.2	3.2	2.8	2.8	2.6
11 Denmark	1.2	1.2	4.6	4.8	3.5	2.2	0.8	0.4	0.3	1.3	1.5	2.2	0.4	1.2	2.1	0.3	1.8	2.1	2.1	1.7	1.7	1.4
12 Finland	1.6	1.6	5.1	5.2	3.9	2.6	1.2	0.8	0.7	1.7	1.9	2.6	0.8	1.6	2.5	0.7	2.2	2.5	2.5	2.1	2.1	1.8
13 France	3.4	3.4	6.9	7.1	5.8	4.5	3.0	2.6	2.5	3.5	3.7	4.5	2.6	3.4	4.3	2.5	4.1	4.3	4.3	3.9	3.9	3.6
14 UK	2.1	2.1	5.6	5.7	4.5	3.2	1.7	1.3	1.3	2.2	2.4	3.2	1.3	2.1	3.0	1.3	2.8	3.0	3.0	2.6	2.6	2.4
15 Ireland	4.5	4.5	8.1	8.2	6.9	5.6	4.1	3.7	3.6	4.6	4.8	5.6	3.7	4.5	5.4	3.6	5.2	5.4	5.4	5.0	5.0	4.8
16 Italy	4.5	4.5	8.1	8.2	6.9	5.6	4.1	3.7	3.6	4.6	4.8	5.6	3.7	4.5	5.4	3.6	5.2	5.5	5.4	5.0	5.0	4.8
17 Netherl	1.2	1.1	4.6	4.7	3.5	2.2	0.7	0.3	0.3	1.2	1.4	2.2	0.4	1.1	2.0	0.3	1.8	2.0	2.0	1.6	1.6	1.4
18 Portugal	1.3	1.3	4.7	4.9	3.6	2.3	0.9	0.5	0.4	1.4	1.6	2.3	0.5	1.3	2.2	0.4	1.9	2.2	2.2	1.8	1.8	1.5
19 Sweden	1.1	1.1	4.6	4.7	3.5	2.2	0.7	0.3	0.3	1.2	1.4	2.2	0.3	1.1	2.0	0.3	1.8	2.0	2.0	1.6	1.6	1.4
20 Eur	1.3	1.3	4.8	4.9	3.6	2.4	0.9	0.5	0.4	1.4	1.6	2.3	0.5	1.3	2.2	0.5	1.9	2.2	2.2	1.8	1.8	1.5
21 Turkey	6.2	6.1	9.8	9.9	8.6	7.3	5.7	5.3	5.3	6.3	6.5	7.3	5.3	6.1	7.1	5.3	6.8	7.1	7.1	6.7	6.7	6.4
22 ROW	3.3	3.2	6.8	6.9	5.6	4.3	2.8	2.4	2.4	3.3	3.5	4.3	2.4	3.2	4.1	2.4	3.9	4.2	4.2	3.8	3.8	3.5

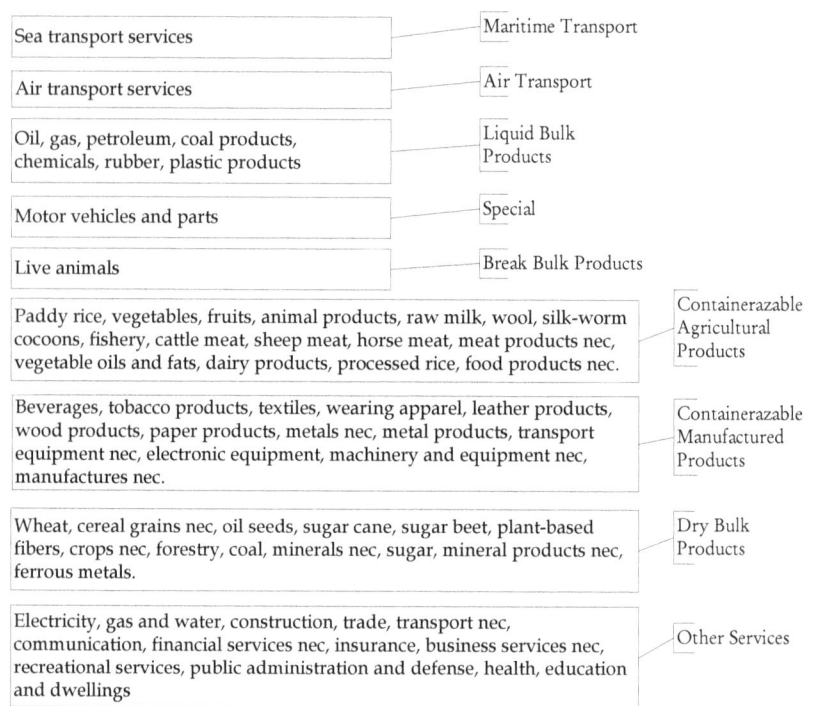

Figure 1. Commodities and sample aggregation (Source: GTAP Database)

Figure 1 depicts a sample aggregation of the commodities to represent their corresponding logistics services.

4. Conclusion

This study looked into the effects on logistics services based an economies in long-term. The results indicate that various logistics services from/to economies exhibit increases or decreases. These results are reported by using dynamic (CGE) simulations.

References

Akune, Y., Okiyama, M., & Tokunaga, S. (2015). Economic Evaluation of Dissemination of High Temperature-Tolerant Rice in Japan Using a Dynamic Computable General Equilibrium Model. *Jarq-Japan Agricultural Research Quarterly, 49*(2), 127-133.

Arndt, C., Pauw, K., & Thurlow, J. (2012). Biofuels and economic development: A computable general equilibrium analysis for Tanzania. *Energy Economics, 34*(6), 1922-1930. doi: 10.1016/j.eneco.2012.07.020

Asafu-Adjaye, J., & Mahadevan, R. (2013). Implications of CO2 reduction policies for a high carbon emitting economy. *Energy Economics, 38*, 32-41. doi: 10.1016/j.eneco.2013.03.004

Bao, Q., Tang, L., Zhang, Z. X., & Wang, S. Y. (2013). Impacts of border carbon adjustments on China's sectoral emissions: Simulations with a dynamic computable general equilibrium model. *China Economic Review, 24*, 77-94. doi: 10.1016/j.chieco.2012.11.002

Barkhordar, Z. A., & Saboohi, Y. (2013). Assessing alternative options for allocating oil revenue in

Iran. *Energy Policy, 63,* 1207-1216. doi: 10.1016/j.enpol.2013.08.099

Berrittella, M., & Zhang, J. (2015). Fiscal sustainability in the EU: From the short-term risk to the long-term challenge. *Journal of Policy Modeling, 37*(2), 261-280. doi: 10.1016/j.jpolmod.2015.02.004

Bhattarai, K. (2015). Financial deepening and economic growth. *Applied Economics, 47*(11), 1133-1150. doi: 10.1080/00036846.2014.993130

Bhattarai, K., & Dixon, H. (2014). Equilibrium Unemployment in a General Equilibrium Model with Taxes. *Manchester School, 82,* 90-128. doi: 10.1111/manc.12066

Boccanfuso, D., Joanis, M., Richard, P., & Savard, L. (2014). A comparative analysis of funding schemes for public infrastructure spending in Quebec. *Applied Economics, 46*(22), 2653-2664. doi: 10.1080/00036846.2014.909576

Bourne, M., Childs, J., & Philippidis, G. (2012). Reaping what others have sown: Measuring the Impact of the global financial crisis on Spanish agriculture. *Itea-Informacion Tecnica Economica Agraria, 108*(4), 405-425.

Breisinger, C., & Ecker, O. (2014). Simulating economic growth effects on food and nutrition security in Yemen: A new macro-micro modeling approach. *Economic Modelling, 43*, 100-113. doi: 10.1016/j.econmod.2014.07.029

Breisinger, C., Engelke, W., & Ecker, O. (2012). Leveraging Fuel Subsidy Reform for Transition in Yemen. *Sustainability, 4*(11), 2862-2887. doi: 10.3390/su4112862

Cheong, I., & Tongzon, J. (2013). Comparing the Economic Impact of the Trans-Pacific Partnership and the Regional Comprehensive Economic Partnership. *Asian Economic Papers, 12*(2), 144-164. doi: 10.1162/ASEP_a_00218

Chi, Y. Y., Guo, Z. Q., Zheng, Y. H., & Zhang, X. P. (2014). Scenarios Analysis of the Energies' Consumption and Carbon Emissions in China Based on a Dynamic CGE Model. *Sustainability, 6*(2), 487-512. doi: 10.3390/su6020487

Cockburn, J., Emini, A. C., & Tiberti, L. (2014). Impacts of the global economic crisis and national policy responses on children in Cameroon. *Canadian Journal of Development Studies-Revue*

Canadienne D Etudes Du Developpement, 35(3), 396-418. doi: 10.1080/02255189.2014.934212

Dai, H. C., Masui, T., Matsuoka, Y., & Fujimori, S. (2012). The impacts of China's household consumption expenditure patterns on energy demand and carbon emissions towards 2050. *Energy Policy, 50*, 736-750. doi: 10.1016/j.enpol.2012.08.023

Decreux, Y., & Fontagne, L. (2015). What Next for Multilateral Trade Talks? Quantifying the Role of Negotiation Modalities. *World Trade Review, 14*(1), 29-43. doi: 10.1017/s1474745614000354

Doumax, V., Philip, J. M., & Sarasa, C. (2014). Biofuels, tax policies and oil prices in France: Insights from a dynamic CGE model. *Energy Policy, 66*, 603-614. doi: 10.1016/j.enpol.2013.11.027

Faehn, T., & Bruvoll, A. (2009). Richer and cleaner-At others' expense? *Resource and Energy Economics, 31*(2), 103-122. doi: 10.1016/j.reseneeco.2008.11.001

Femenia, F. (2010). Impacts of Stockholding Behaviour on Agricultural Market Volatility: A Dynamic Computable General Equilibrium

Approach. *German Journal of Agricultural Economics, 59*(3), 187-201.

Femenia, F., & Gohin, A. (2013). On the optimal implementation of agricultural policy reforms. *Journal of Policy Modeling, 35*(1), 61-74. doi: 10.1016/j.jpolmod.2012.05.019

Furuya, J., Tokunaga, S., Okiyama, M., Akune, Y., Kunimitsu, Y., Aizaki, H., & Kobayashi, S. (2015). Economic Evaluation of Agricultural Mitigation and Adaptation Technologies for Climate Change: Model Development for Impact Analysis and Technological Assessment. *Jarq-Japan Agricultural Research Quarterly, 49*(2), 119-125.

Gohin, A., & Rault, A. (2013). Assessing the economic costs of a foot and mouth disease outbreak on Brittany: A dynamic computable general equilibrium analysis. *Food Policy, 39*, 97-107. doi: 10.1016/j.foodpol.2013.01.003

He, J., Chen, X. K., Shi, Y., & Li, A. H. (2007). Dynamic computable general equilibrium model and sensitivity analysis for shadow price of water resource in China. *Water Resources Management, 21*(9), 1517-1533. doi: 10.1007/s11269-006-9102-7

Hosoe, N. (2014). Japanese manufacturing facing post-Fukushima power crisis: a dynamic computable general equilibrium analysis with foreign direct investment. *Applied Economics, 46*(17), 2010-2020. doi: 10.1080/00036846.2014.892198

Jiang, X. M., & Mai, Y. H. (2015). The social welfare housing project and its effects in China. *Journal of Systems Science & Complexity, 28*(2), 393-408. doi: 10.1007/s11424-014-3261-z

Lakatos, C., & Walmsley, T. (2012). Investment creation and diversion effects of the ASEAN-China free trade agreement. *Economic Modelling, 29*(3), 766-779. doi: 10.1016/j.econmod.2012.02.004

Lanzi, E., Chateau, J., & Dellink, R. (2012). Alternative approaches for levelling carbon prices in a world with fragmented carbon markets. *Energy Economics, 34,* S240-S250. doi: 10.1016/j.eneco.2012.08.016

Li, N., Wang, X. J., Shi, M. J., & Yang, H. (2015). Economic Impacts of Total Water Use Control in the Heihe River Basin in Northwestern China-An Integrated CGE-BEM Modeling Approach.

Sustainability, 7(3), 3460-3478. doi: 10.3390/su7033460

Liang, Q. M., Wang, Q., & Wei, Y. M. (2013). Assessing the Distributional Impacts of Carbon Tax among Households across Different Income Groups: The Case of China. *Energy & Environment, 24*(7-8), 1323-1346.

Liang, Q. M., Yao, Y. F., Zhao, L. T., Wang, C., Yang, R. G., & Wei, Y. M. (2014). Platform for China Energy & Environmental Policy Analysis: A general design and its application. *Environmental Modelling & Software, 51*, 195-206. doi: 10.1016/j.envsoft.2013.09.032

Liu, Y., & Lu, Y. Y. (2015). The Economic impact of different carbon tax revenue recycling schemes in China: A model-based scenario analysis. *Applied Energy, 141*, 96-105. doi: 10.1016/j.apenergy.2014.12.032

Mabugu, R. E., Fofana, I., & Chitiga-Mabugu, M. R. (2015). Pro-Poor Tax Policy Changes in South Africa: Potential and Limitations. *Journal of African Economies, 24*, II73-II105. doi: 10.1093/jae/eju038

Mai, Y. H., Peng, X. J., Dixon, P., & Rimmer, M. (2014). The economic effects of facilitating the flow of rural workers to urban employment in China. *Papers in Regional Science, 93*(3), 619-642. doi: 10.1111/pirs.12004

Mariano, M. J. M., Giesecke, J. A., & Tran, N. H. (2015). The effects of domestic rice market interventions outside business-as-usual conditions for imported rice prices. *Applied Economics, 47*(8), 809-832. doi: 10.1080/00036846.2014.980576

Matovu, J. M. (2012). Trade Reforms and Horizontal Inequalities: The Case of Uganda. *European Journal of Development Research, 24*(5), 753-776. doi: 10.1057/ejdr.2012.35

Parrado, R., & De Cian, E. (2014). Technology spillovers embodied in international trade: Intertemporal, regional and sectoral effects in a global CGE framework. *Energy Economics, 41,* 76-89. doi: 10.1016/j.eneco.2013.10.016

Philip, J. M., Sanchez-Choliz, J., & Sarasa, C. (2014). Technological change in irrigated agriculture in a semiarid region of Spain. *Water Resources*

Research, 50(12), 9221-9235. doi: 10.1002/2014wr015728

Philippidis, G., & Hubbard, L. (2005). A dynamic computable general equilibrium treatment of the ban on UK beef exports: A note. *Journal of Agricultural Economics, 56*(2), 307-312. doi: 10.1111/j.1477-9552.2005.00006.x

Qin, C. B., Bressers, H. T. A., Su, Z., Jia, Y. W., & Wang, H. (2011). Assessing economic impacts of China's water pollution mitigation measures through a dynamic computable general equilibrium analysis. *Environmental Research Letters, 6*(4), 15. doi: 10.1088/1748-9326/6/4/044026

Ricci, O. (2012). Providing adequate economic incentives for bioenergies with CO2 capture and geological storage. *Energy Policy, 44*, 362-373. doi: 10.1016/j.enpol.2012.01.066

Ruamsuke, K., Dhakar, S., & Marpaung, C. O. P. (2015). Energy and economic impacts of the global climate change policy on Southeast Asian countries: A general equilibrium analysis. *Energy, 81*, 446-461. doi: 10.1016/j.energy.2014.12.057

Saveyn, B., Paroussos, L., & Ciscar, J. C. (2012). Economic analysis of a low carbon path to 2050: A case for China, India and Japan. *Energy Economics, 34,* S451-S458. doi: 10.1016/j.eneco.2012.04.010

Schenker, O. (2013). Exchanging Goods and Damages: The Role of Trade on the Distribution of Climate Change Costs. *Environmental & Resource Economics, 54*(2), 261-282. doi: 10.1007/s10640-012-9593-z

Seung, C. K., & Kraybill, D. S. (2001). The effects of infrastructure investment: A two-sector dynamic computable general equilibrium analysis for Ohio. *International Regional Science Review, 24*(2), 261-281. doi: 10.1177/016001701761013150

Verikios, G., Dixon, P. B., Rimmer, M. T., & Harris, A. H. (2015). Improving health in an advanced economy: An economywide analysis for Australia. *Economic Modelling, 46,* 250-261. doi: 10.1016/j.econmod.2014.12.032

Wittwer, G., & Banerjee, O. (2015). Investing in irrigation development in North West Queensland, Australia. *Australian Journal of*

Agricultural and Resource Economics, 59(2), 189-207. doi: 10.1111/1467-8489.12057

Wu, T., Zhang, M. B., & Ou, X. M. (2014). Analysis of Future Vehicle Energy Demand in China Based on a Gompertz Function Method and Computable General Equilibrium Model. *Energies, 7*(11), 7454-7482. doi: 10.3390/en7117454

Xie, W., Li, N., Wu, J. D., & Hao, X. L. (2014). Modeling the economic costs of disasters and recovery: analysis using a dynamic computable general equilibrium model. *Natural Hazards and Earth System Sciences, 14*(4), 757-772. doi: 10.5194/nhess-14-757-2014

Xie, W., Li, N., Wu, J. D., & Hao, X. L. (2015). Disaster Risk Decision: A Dynamic Computable General Equilibrium Analysis of Regional Mitigation Investment. *Human and Ecological Risk Assessment, 21*(1), 81-99. doi: 10.1080/10807039.2013.871997

Appendix A - Preliminaries

This content is adapted from the dynamic GTAP source code.

For more documentation, refer to:
- Hertel, T.W. and M.E. Tsigas "Structure of the Standard GTAP Model", Chapter 2 in T.W. Hertel (editor) *Global Trade Analysis: Modeling and Applications*, Cambridge University Press, 1997.
- Ianchovichina, E.I. (1998) *International Capital Linkages: Theory and Applications in A Dynamic Computable General Equilibrium Model*.
- Ianchovichina, E.I. and R. McDougall, "Theoretical Structure of Dynamic GTAP", *GTAP Technical Paper No. 17*, Dec. 2000

Preliminaries:
- **Files**

File	
GTAPSETS	# file with set specification #;
GTAPDATA	# file containing all base data #;
GTAPPARM	# file containing behavioral parameters #;
GTAPPARMK	# special parameters for dynamics #;

- **Sets**

Sets define relevant groupings of entities over which we will be performing operations in the model. SUBSETS are defined in order to facilitate summation over only a portion of a given group, e.g., tradable commodities.

Aide-Memoire for Sets

	DEMD_COMM		
ENDW_COMM	TRAD_COMM		CGDS_COMM
	NSAV_COMM		
		PROD_COMM	

For Endowments,

ENDW_COMM	
ENDWM_COMM	ENDWS_COMM

- **"Read" statements of Base Data**

We read in here almost all the base data, and define variables and coefficients associated with them. A few data arrays used each in a single module are read in those modules: VKB, VTMFSD, and DPARSUM.

The READ statements are divided into six sections:

1. Saving
2. Government Consumption
3. Private Consumption
4. Firms
5. Global Trust
6. International Trade and Transport
7. Regional Household

Since these are invariant for each solution of the model, they are termed coefficients. Coefficients are assigned upper case to distinguish them from variables. (This is purely cosmetic, as GEMPACK is not case-sensitive.) Variables in GEMPACK are assigned lower case labels to
denote the fact that they are percentage changes. In some cases, original levels values for selected variables are defined to permit the user to compare post-simulation levels values across several simulations.

The updating command indicates how the new level of the coefficient will be computed based on the previous solution of the linearized equations. Note that the notation used in the update commands is a shorthand for total differentials of these coefficient values. Thus, w * v indicates that we want to take the total differential of W * V, plug in the calculated values of w and v, and add this to the base level in order to obtain a revised value for this product.

- **Common "Variables"**

Common variables are defined as variables which are used in more than one module. For example, the variable y(r) is used in the Government Consumption, Private Consumption, Firms, Regional Household and Investment, Global Bank and Savings modules. Appendices, e.g., Summary

Indices, are not included in this definition.

- **Common "Coefficients"**
- **Key Derivatives of the Base Data**
- **Regional Expenditure and Income**

Regional income is allocated between private consumption expenditure, government consumption expenditure, and saving.

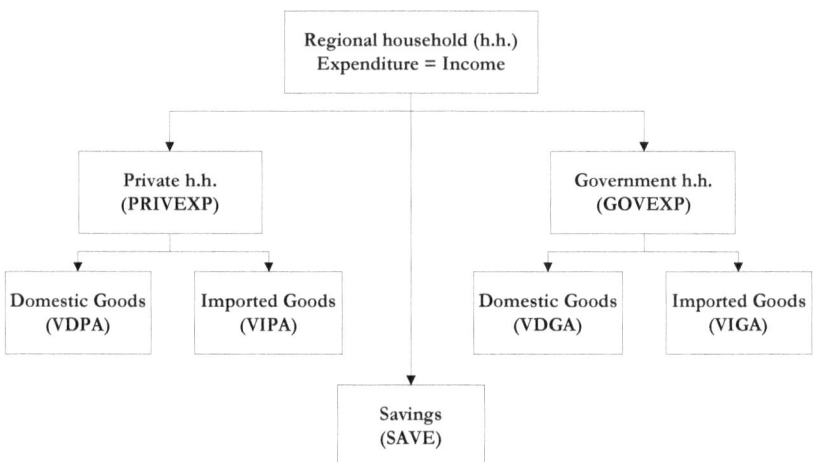

Figure. Expenditure of Regional Household

Note: The coefficients at the ends of branches are Base Data, e.g., VDPA, SAVE.

- **Indirect Tax Receipts**
- **Miscellaneous Coefficients**

Common Coefficients are defined as coefficients, which are used in more than one module. For example, ESUBD(i) is used in the Government Household, Private Household, and Firms modules.

Modules:
1. Government Consumption
2. Private Consumption
3. Firms

We now turn to the behavioral equations for firms. The following picture describes factor demands. The first set of equations describe demands for primary factors. (See table 4 of Hertel and Tsigas.)

Production structure

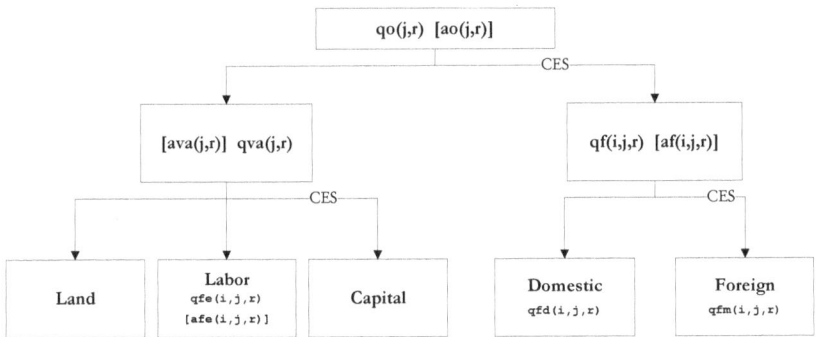

Figure. Production structure

4. Physical Capital, Global Trust, and Savings

Dynamics extension: investment theory

These equations determine investment by region, `qcgds'. They also determine descriptive variables representing regional rates of return (`DROR') and the world average rate of return (`DRORW').

That's right, the rate of return variable is now merely descriptive. Investment responds to the expected not the actual rate. The actual rates does indirectly affect behavior, via an expectations adjustment process. But as it happens, it appears there as a coefficient not as a variable.

Another change: the rate of return variable is now an absolute change variable. This is appropriate since the (net) rate of return may be either positive or negative. We also include variables pertaining to (expected and target) gross rates of return: these are percentage change variables, because the gross rate of return is always positive; because however high the stock of capital relative to the demand, there is always some strictly positive capital rental low enough to
clear the market.

Some change in nomenclature: we use `qk' for capital stock, not `kb', because we do not now have separate variables for beginning- and end-of-period capital stocks, because we don't have a period, just a point in time. We use `qk' not 'k' because `k' appears elsewhere as an index. Also, we use `VK' for the capital stock coefficient, not `VKB'.

5. International Trade
6. International Transport Services
7. Regional Household
8. Equilibrium Conditions

Appendices:
A. Summary Indices
B. Equivalent Variation
C. Welfare Decomposition
D. Terms of Trade Decomposition

Preliminaries

Files

File	
GTAPSETS	# file with set specification #;
GTAPDATA	# file containing all base data #;
GTAPPARM	# file containing behavioral parameters #;
GTAPPARMK	# special parameters for dynamics #;

Sets

Sets define relevant groupings of entities over which we will be performing operations in the model. SUBSETS are defined in order to facilitate summation over only a portion of a given group, e.g., tradeable commodities.

Aide-Memoire for Sets

	DEMD_COMM	
ENDW_COMM	TRAD_COMM	CGDS_COMM
	NSAV_COMM	
	PROD_COMM	

For Endowments,

ENDW_COMM	
ENDWM_COMM	ENDWS_COMM

Coefficient
VERNUM # version of GTAP data #;
Read
VERNUM from file GTAPDATA header "DVER";
Update (change)
VERNUM = 0.0; ! force it to be copied to update file !

Set
REG # regions in the model #
maximum size 10 read elements from file GTAPSETS header "H1";

Set
TRAD_COMM # traded commodities #
maximum size 10 read elements from file GTAPSETS header "H2";

Set
MARG_COMM # margin commodities #
maximum size 10 read elements from file GTAPSETS header "MARG";

Subset
MARG_COMM is subset of TRAD_COMM;

Set
NMRG_COMM # non-margin commodities # = TRAD_COMM -

```
MARG_COMM;
```

```
Set
CGDS_COMM # capital goods commodities #
maximum size 1 read elements from file GTAPSETS header "H9";
```

```
Set
ENDW_COMM # endowment commodities #
maximum size 5 read elements from file GTAPSETS header "H6";
```

```
Set
ENDWC_COMM # capital endowment commodity #
maximum size 1 read elements from file GTAPSETS header "HD0";
```

```
Subset
ENDWC_COMM is subset of ENDW_COMM;
```

```
Set
PROD_COMM # produced commodities # = TRAD_COMM union CGDS_COMM;
```

```
Set
DEMD_COMM # demanded commodities # = ENDW_COMM union TRAD_COMM;
```

```
Set
NSAV_COMM # non-savings commodities # = DEMD_COMM union CGDS_COMM;
```

```
Set
ENDWNA_COMM
# non-accumulable endowment commodities # = ENDW_COMM - ENDWC_COMM;
```

```
Subset
PROD_COMM is subset of NSAV_COMM;
```

Check for non-overlapping sets (users can't use the same name for elements of ENDW_COMM,

TRAD_COMM or CGDS_COMM).

Coefficient
SIZE_TRAD # size of TRAD_COMM set #;
Formula
SIZE_TRAD = sum(i,TRAD_COMM, 1);

Coefficient
SIZE_ENDW # size of ENDW_COMM set #;
Formula
SIZE_ENDW = sum(i,ENDW_COMM, 1);

Coefficient
SIZE_DEMD # size of DEMD_COMM set #;
Formula
SIZE_DEMD = sum(i,DEMD_COMM, 1);
Assertion (initial)
SIZE_DEMD = SIZE_TRAD + SIZE_ENDW;

Coefficient
SIZE_CGDS # size of CGDS_COMM set #;
Formula
SIZE_CGDS = sum(i,CGDS_COMM, 1);

Coefficient
SIZE_PROD # size of PROD_COMM set #;
Formula
SIZE_PROD = sum(i,PROD_COMM, 1);
Assertion (initial)
SIZE_PROD = SIZE_TRAD + SIZE_CGDS;

The sluggish endowments are now defined dynamically, based on the variable SLUG. SLUG is a binary variable that is zero for mobile endowments and one for sluggish endowments. We must define and read in this variable before proceeding further.

Coefficient (integer,parameter) (all,i,ENDW_COMM)
SLUG(i) # sluggish primary factor endowments #;
Read
SLUG from file GTAPPARM header "SLUG";

Set
ENDWS_COMM
sluggish endowment commodities # = (all,i,ENDW_COMM: SLUG(i)>0);

Set
ENDWM_COMM
mobile endowment commodities # = ENDW_COMM - ENDWS_COMM;

"Read" Statements of Base Data

We read in here almost all the base data, and define variables and coefficients associated with them. A few data arrays used each in a single module are read in those modules: VKB, VTMFSD, and DPARSUM.

The READ statements are divided into six sections:

    8. Saving
    9. Government Consumption
    10. Private Consumption
    11. Firms
    12. Global Trust
    13. International Trade and Transport
    14. Regional Household

Since these are invariant for each solution of the model, they are termed coefficients. Coefficients are assigned upper case to distinguish them from variables. (This is purely cosmetic, as GEMPACK is not case-sensitive.) Variables in GEMPACK are assigned lower case labels to
denote the fact that they are percentage changes. In some cases, original levels values for selected variables are defined to permit the user to compare post-simulation levels values across several simulations.

The updating command indicates how the new

level of the coefficient will be computed based on the previous solution of the linearized equations. Note that the notation used in the update commands is a shorthand for total differentials of these coefficient values. Thus, w * v indicates that we want to take the total differential of W * V, plug in the calculated values of w and v, and add this to the base level in order to obtain a revised value for this product.

Saving

Variable (all,r,REG)
psave(r) # price of savings in region r #;

Variable (all,r,REG)
qsave(r) # regional demand for NET savings #;

Coefficient (all,r,REG)
SAVE(r) # expenditure on NET savings in region r valued at agent's prices #;

Update (all,r,REG)
SAVE(r) = psave(r) * qsave(r);

Read
SAVE from file GTAPDATA header "SAVE";

Government Consumption

Variable (all,i,TRAD_COMM)(all,s,REG)
pgd(i,s) # price of domestic i in government consumption in s #;

Variable (orig_level=VDGM)(all,i,TRAD_COMM)(all,s,REG)
qgd(i,s) # government hhld demand for domestic i in region s #;

Coefficient (ge 0)(all,i,TRAD_COMM)(all,r,REG)
VDGA(i,r) # government consumption expenditure on domestic i in r #;

Update (all,i,TRAD_COMM)(all,r,REG)
VDGA(i,r) = pgd(i,r) * qgd(i,r);

Read
VDGA from file GTAPDATA header "VDGA";

Variable (orig_level=1.0)(all,i,NSAV_COMM)(all,r,REG)
pm(i,r) # market price of commodity i in region r #;

Coefficient (ge 0)(all,i,TRAD_COMM)(all,r,REG)
VDGM(i,r) # government consumption expenditure on domestic i in r #;

Update (all,i,TRAD_COMM)(all,r,REG)
VDGM(i,r) = pm(i,r) * qgd(i,r);

Read
VDGM from file GTAPDATA header "VDGM";

Variable (all,i,TRAD_COMM)(all,s,REG)
pgm(i,s) # price of imports of i in government consumption in s #;

Variable (orig_level=VIGM)(all,i,TRAD_COMM)(all,s,REG)
qgm(i,s) # government hhld demand for imports of i in region s #;

Coefficient (ge 0)(all,i,TRAD_COMM)(all,r,REG)
VIGA(i,r) # government consumption expenditure on imported i #;

Update (all,i,TRAD_COMM)(all,r,REG)

VIGA(i,r) = pgm(i,r) * qgm(i,r);

Read
VIGA from file GTAPDATA header "VIGA";

Variable (orig_level=1.0)(all,i,TRAD_COMM)(all,r,REG)
pim(i,r) # market price of composite import i in region r #;

Coefficient (ge 0)(all,i,TRAD_COMM)(all,r,REG)
VIGM(i,r) # gov't consumption expenditure on i in r #;

Update (all,i,TRAD_COMM)(all,r,REG)
VIGM(i,r) = pim(i,r) * qgm(i,r);

Read
VIGM from file GTAPDATA header "VIGM";

Private Consumption

```
Variable (all,i,TRAD_COMM)(all,s,REG)
ppd(i,s) # price of domestic i to private households in s #;
```

```
Variable (orig_level=VDPM)(all,i,TRAD_COMM)(all,s,REG)
qpd(i,s) # private hhld demand for domestic i in region s #;
```

```
Coefficient (ge 0)(all,i,TRAD_COMM)(all,r,REG)
VDPA(i,r) # private consumption expenditure on domestic i in r #;
```

```
Update (all,i,TRAD_COMM)(all,r,REG)
VDPA(i,r) = ppd(i,r) * qpd(i,r);
```

```
Read
VDPA from file GTAPDATA header "VDPA";
```

```
Coefficient (ge 0)(all,i,TRAD_COMM)(all,r,REG)
VDPM(i,r) # private consumption expenditure on domestic i in r #;
```

```
Update (all,i,TRAD_COMM)(all,r,REG)
VDPM(i,r) = pm(i,r) * qpd(i,r);
```

```
Read
VDPM from file GTAPDATA header "VDPM";
```

```
Variable (all,i,TRAD_COMM)(all,s,REG)
ppm(i,s) # price of imports of i by private households in s #;
```

```
Variable (orig_level=VIPM)(all,i,TRAD_COMM)(all,s,REG)
qpm(i,s) # private hhld demand for imports of i in region s #;
```

```
Coefficient (ge 0)(all,i,TRAD_COMM)(all,r,REG)
VIPA(i,r) # private consumption expenditure on imported i in r #;
```

```
Update (all,i,TRAD_COMM)(all,r,REG)
VIPA(i,r) = ppm(i,r) * qpm(i,r);
```

```
Read
VIPA from file GTAPDATA header "VIPA";
```

Coefficient (ge 0)(all,i,TRAD_COMM)(all,r,REG)
VIPM(i,r) # private consumption expenditure on i in r #;

Update (all,i,TRAD_COMM)(all,r,REG)
VIPM(i,r)= pim(i,r) * qpm(i,r);

Read
VIPM from file GTAPDATA header "VIPM";

Firms

Variable (all,i,NSAV_COMM)(all,r,REG)
ps(i,r) # supply price of commodity i in region r #;

Variable (orig_level=VOM)(all,i,NSAV_COMM)(all,r,REG)
qo(i,r) # industry output of commodity i in region r #;

Coefficient (ge 0)(all,i,ENDW_COMM)(all,r,REG)
EVOA(i,r) # value of commodity i output in region r #;

Update (all,i,ENDW_COMM)(all,r,REG)
EVOA(i,r) = ps(i,r) * qo(i,r);

Read
EVOA from file GTAPDATA header "EVOA";

Variable (all,i,ENDW_COMM)(all,j,PROD_COMM)(all,r,REG)
pfe(i,j,r) # firms' price for endowment commodity i in ind. j, region r #;

Variable (orig_level=VFM)(all,i,ENDW_COMM)(all,j,PROD_COMM)(all,r,REG)
qfe(i,j,r) # demand for endowment i for use in ind. j in region r #;

Coefficient (ge 0)(all,i,ENDW_COMM)(all,j,PROD_COMM)(all,r,REG)
EVFA(i,j,r) # producer expenditure on i by j in r at agent's prices #;

Update (all,i,ENDW_COMM)(all,j,PROD_COMM)(all,r,REG)
EVFA(i,j,r) = pfe(i,j,r) * qfe(i,j,r);

Read
EVFA from file GTAPDATA header "EVFA";

Variable (all,i,TRAD_COMM)(all,j,PROD_COMM)(all,s,REG)
pfd(i,j,s) # price index for domestic purchases of i by j in region s #;

Variable (orig_level=VDFM)(all,i,TRAD_COMM)(all,j,PROD_COMM)(all,s,REG)
qfd(i,j,s) # domestic good i demanded by industry j in region s #;

Coefficient (ge 0)(all,i,TRAD_COMM)(all,j,PROD_COMM)(all,r,REG)
VDFA(i,j,r) # purchases of domestic i for use by j in region r #;

Update (all,i,TRAD_COMM)(all,j,PROD_COMM)(all,r,REG)
VDFA(i,j,r) = pfd(i,j,r) * qfd(i,j,r);

Read
VDFA from file GTAPDATA header "VDFA";

Variable (all,i,TRAD_COMM)(all,j,PROD_COMM)(all,s,REG)
pfm(i,j,s) # price index for imports of i by j in region s #;

Variable (orig_level=VIFM)(all,i,TRAD_COMM)(all,j,PROD_COMM)(all,s,REG)
qfm(i,j,s) # demand for i by industry j in region s #;

Coefficient (ge 0)(all,i,TRAD_COMM)(all,j,PROD_COMM)(all,r,REG)
VIFA(i,j,r) # purchases of imported i for use by j in region r #;

Update (all,i,TRAD_COMM)(all,j,PROD_COMM)(all,r,REG)
VIFA(i,j,r) = pfm(i,j,r) * qfm(i,j,r);

Read
VIFA from file GTAPDATA header "VIFA";

Variable (all,i,ENDWS_COMM)(all,j,PROD_COMM)(all,r,REG)
pmes(i,j,r) # market price of sluggish endowment i used by j in r #;

Coefficient (ge 0)(all,i,ENDW_COMM)(all,j,PROD_COMM)(all,r,REG)
VFM(i,j,r) # producer expenditure on i by j in r valued at mkt prices #;

Update (all,i,ENDWM_COMM)(all,j,PROD_COMM)(all,r,REG)
VFM(i,j,r) = pm(i,r) * qfe(i,j,r);

Update (all,i,ENDWS_COMM)(all,j,PROD_COMM)(all,r,REG)
VFM(i,j,r) = pmes(i,j,r) * qfe(i,j,r);

Read
VFM from file GTAPDATA header "VFM";

Coefficient (ge 0)(all,i,TRAD_COMM)(all,j,PROD_COMM)(all,r,REG)
VIFM(i,j,r) # purchases of imports i for use by j in region r #;

Update (all,i,TRAD_COMM)(all,j,PROD_COMM)(all,r,REG)
VIFM(i,j,r) = pim(i,r) * qfm(i,j,r);

Read
VIFM from file GTAPDATA header "VIFM";

Coefficient (ge 0)(all,i,TRAD_COMM)(all,j,PROD_COMM)(all,r,REG)
VDFM(i,j,r) # purchases of domestic i for use by j in region r #;

Update (all,i,TRAD_COMM)(all,j,PROD_COMM)(all,r,REG)
VDFM(i,j,r) = pm(i,r) * qfd(i,j,r);

Read
VDFM from file GTAPDATA header "VDFM";

Global Trust

Variable (all,r,REG)
qk(r) # Capital stock located in region r #;

Variable (all,r,REG)
pcgds(r) # price of investment goods = ps("cgds",r) #;

Coefficient (ge 0)(all,r,REG)
VDEP(r) # value of capital depeciation in r (exogenous) #;

Update (all,r,REG)
VDEP(r) = qk(r) * pcgds(r);

Read
VDEP from file GTAPDATA header "VDEP";

Coefficient (ge 0)(all,r,REG)
VK(r) # value of capital stock, in region r #;

Update (all,r,REG)
VK(r) = qk(r) * pcgds(r);

Read
VK from file GTAPDATA header "VKB";

Variable (all,r,REG)
yqtf(r) # global trust income from equity in firms in region r #;

Coefficient (ge 0)(all,r,REG)
YQTFIRM(r) # income of global trust from firms in region r #;

Update (all,r,REG)
YQTFIRM(r) = yqtf(r);

Read
YQTFIRM from file GTAPDATA header "YQTF";

International Trade and Transport

Variable (all,i,TRAD_COMM)(all,r,REG)(all,s,REG)
pms(i,r,s) # domestic price for good i supplied from r to region s #;

Variable (orig_level=VXMD)(all,i,TRAD_COMM)(all,r,REG)(all,s,REG)
qxs(i,r,s) # export sales of commodity i from r to region s #;

Coefficient (ge 0)(all,i,TRAD_COMM)(all,r,REG)(all,s,REG)
VIMS(i,r,s) # imports of i from r to s valued at domestic mkt prices #;

Update (all,i,TRAD_COMM)(all,r,REG)(all,s,REG)
VIMS(i,r,s) = pms(i,r,s) * qxs(i,r,s);

Read
VIMS from file GTAPDATA header "VIMS";

Variable (all,i,TRAD_COMM)(all,r,REG)(all,s,REG)
pcif(i,r,s) # CIF world price of commodity i supplied from r to s #;

Coefficient (ge 0)(all,i,TRAD_COMM)(all,r,REG)(all,s,REG)
VIWS(i,r,s) # imports of i from r to s valued CIF (tradeables only) #;

Update (all,i,TRAD_COMM)(all,r,REG)(all,s,REG)
VIWS(i,r,s) = pcif(i,r,s) * qxs(i,r,s);

Read
VIWS from file GTAPDATA header "VIWS";

Variable (all,i,TRAD_COMM)(all,r,REG)(all,s,REG)
pfob(i,r,s) # FOB world price of commodity i supplied from r to s #;

Coefficient (ge 0)(all,i,TRAD_COMM)(all,r,REG)(all,s,REG)
VXWD(i,r,s) # exports of i from r to s valued FOB (tradeables only) #;

Update (all,i,TRAD_COMM)(all,r,REG)(all,s,REG)
VXWD(i,r,s) = pfob(i,r,s) * qxs(i,r,s);

Read
VXWD from file GTAPDATA header "VXWD";

Coefficient (ge 0)(all,i,TRAD_COMM)(all,r,REG)(all,s,REG)
VXMD(i,r,s)
# exports of i from r to s valued at mkt prices (tradeables only) #;

Update (all,i,TRAD_COMM)(all,r,REG)(all,s,REG)
VXMD(i,r,s) = pm(i,r) * qxs(i,r,s);

Read
VXMD from file GTAPDATA header "VXMD";

Variable (orig_level=VST)(all,m,MARG_COMM)(all,r,REG)
qst(m,r) # sales of m from r to international transport #;

Coefficient (ge 0)(all,m,MARG_COMM)(all,r,REG)
VST(m,r)
# exprts of m from r for int'l trnsport valued at mkt p (tradeables only) #;

Update (all,m,MARG_COMM)(all,r,REG)
VST(m,r) = pm(m,r) * qst(m,r);

Read
VST from file GTAPDATA header "VST";

Regional Household

Variable (all,r,REG)
yqhf(r) # regional household income from equity in domestic firms #;

Coefficient (ge 0)(all,r,REG)
YQHFIRM(r) # income of region r from local firms #;

Update (all,r,REG)
YQHFIRM(r) = yqhf(r);

Read
YQHFIRM from file GTAPDATA header "YQHF";

Variable (all,r,REG)
yqht(r) # regional household income from equity in the global trust #;

Coefficient (ge 0)(all,r,REG)
YQHTRUST(r) # regional income from global trust #;

Update (all,r,REG)
YQHTRUST(r) = yqht(r);

Read
YQHTRUST from file GTAPDATA header "YQHT";

Common "Variables"

Common variables are defined as variables which are used in more than one module. For example, the variable y(r) is used in the Government Consumption, Private Consumption, Firms, Regional Household and Investment, Global Bank and Savings modules. Appendices, e.g., Summary Indices, are not included in this definition.

Variable (all,r,REG)
y(r) # regional household income in region r #;

Variable (all,r,REG)
pop(r) # regional population #;

Variable (all,i,ENDWS_COMM)(all,j,PROD_COMM)(all,r,REG)
qoes(i,j,r) # supply of sluggish endowment i used by j in r #;

Variable (all,i,ENDW_COMM)(all,r,REG)
endwslack(i,r) # slack variable in endowment market clearing condition #;

This is exogenous, unless the user wishes to employ a partial equilibrium closure in which the price of one or more of the primary factors is fixed.

Variable (all,r,REG)
pgov(r) # price index for gov't hhld expenditure in region r #;

Variable (all,r,REG)
yg(r) # regional government consumption expenditure in region r #;

Variable (all,r,REG)
ug(r) # per capita utility from gov't expend. in region r #;

Variable (all,r,REG)
ppriv(r) # price index for private consumption expenditure in region r #;

Variable (all,r,REG)
uepriv(r) # elasticity of cost wrt utility from private consumption #;

Variable (all,r,REG)
yp(r) # regional private consumption expenditure in region r #;

Variable (all,r,REG)
up(r) # per capita utility from private expend. in region r #;

Variable (all,i,NSAV_COMM)(all,r,REG)
to(i,r) # output (or income) tax in region r #;

Note: It is important that the user NOT shock the tax on capital goods output, as this will cause an inconsistency in the update relationship for VOM(cgds).

Variable (orig_level=VIM)(all,i,TRAD_COMM)(all,s,REG)
qim(i,s) # aggregate imports of i in region s, market price weights #;

Variable (all, r, REG)
qcgds(r) # Output of capital goods sector = qo("cgds",r) #;

Variable (all, r, REG)
qcgds_d(r) # Output of capital goods sector = qo("cgds",r) #;

Variable
yqt # income of the global trust #;

Variable (all,r,REG)
yq_f(r) # income generated by equity / capital in region r #;

Variable
wqt # equity held by the global trust #;

Variable
wq_t # equity held in the global trust #;

Variable (all,r,REG)
wq_f(r) # equity in firms in region r #;

Variable (all,r,REG)
wqhf(r) # equity held by the regional household in domestic firms #;

Variable (all,r,REG)
wqht(r) # equity held by the regional hhld in the global trust #;

Variable (all,r,REG)
wqtf(r) # equity held by the global trust in firms in region r #;

Variable
pqtrust # price of equity in the global trust #;

Variable (change)
time # absolute change in time measured in years #;

The following variables could be dropped when converting to levels equation for income. They are only needed for the linearized equation. The idea here is to look at the ratio of taxes to income in order to preserve homogeneity in prices. (We could also look at changes in tax revenue, but then a uniform price increase would change this variable.) Obviously a simple percentage change variable doesn't work, since many taxes are initially zero. The basic logic of this approach is as follows:

Let R be the ratio of taxes to income: R = T/Y, then:
dR = d(T/Y) = R(t - y)/100
multiply through by Y to get:
YdR = dT - Ty/100
This ratio change is computed for each tax type and for total taxes.

Then the change in tax revenue itself may be computed as:
dT = YdR + Ty/100
in order to determine regional income.

Variable (change) (all,r,REG)
del_taxrgc(r) # change in ratio of government consumption tax to INCOME #;

Variable (change) (all,r,REG)
del_taxrpc(r) # change in ratio of private consumption tax to INCOME #;

Variable (change) (all,r,REG)
del_taxriu(r) # change in ratio of tax on intermediate usage to INCOME #;

Variable (change) (all,r,REG)
del_taxrfu(r) # change in ratio of tax on primary factor usage to INCOME #;

Variable (change) (all,r,REG)
del_taxrout(r) # change in ratio of output tax to INCOME #;

Variable (change) (all,r,REG)
del_taxrexp(r) # change in ratio of export tax to INCOME #;

Variable (change) (all,r,REG)
del_taxrimp(r) # change in ratio of import tax to INCOME #;

Variable (change) (all,r,REG)
del_taxrinc(r) # change in ratio of income tax to INCOME #;

Common "Coefficients"

- **Key Derivatives of the Base Data**
- **Regional Expenditure and Income**
- **Indirect Tax Receipts**
- **Miscellaneous Coefficients**

Common Coefficients are defined as coefficients which are used in more than one module. For example, ESUBD(i) is used in the Government Household, Private Household, and Firms modules.

Key Derivatives of the Base Data

Coefficient (all,i,DEMD_COMM)(all,j,PROD_COMM)(all,r,REG)
VFA(i,j,r) # producer expenditure on i by j in r valued at agent's prices #;

Formula (all,i,ENDW_COMM)(all,j,PROD_COMM)(all,r,REG)
VFA(i,j,r) = EVFA(i,j,r);

Formula (all,i,TRAD_COMM)(all,j,PROD_COMM)(all,s,REG)
VFA(i,j,s) = VDFA(i,j,s) + VIFA(i,j,s);

Coefficient (all,i,NSAV_COMM)(all,r,REG)
VOA(i,r) # value of commodity i output in region r at agent's prices #;

Formula (all,i,ENDW_COMM)(all,r,REG)
VOA(i,r) = EVOA(i,r);
Formula (all,i,PROD_COMM)(all,r,REG)
VOA(i,r) = sum(j,DEMD_COMM, VFA(j,i,r));

Coefficient (all,i,TRAD_COMM)(all,r,REG)
VDM(i,r) # domestic sales of i in r at mkt prices (tradeables only) #;
Formula (all,i,TRAD_COMM)(all,r,REG)
VDM(i,r) = VDPM(i,r) + VDGM(i,r) + sum(j,PROD_COMM, VDFM(i,j,r));

Coefficient (all,i,NSAV_COMM)(all,r,REG)
VOM(i,r) # value of commodity i output in region r at market prices #;
Formula (all,i,ENDW_COMM)(all,r,REG)
VOM(i,r) = sum(j,PROD_COMM, VFM(i,j,r));
Formula (all,m,MARG_COMM)(all,r,REG)
VOM(m,r) = VDM(m,r) + sum(s,REG, VXMD(m,r,s)) + VST(m,r);
Formula (all,i,NMRG_COMM)(all,r,REG)
VOM(i,r) = VDM(i,r) + sum(s,REG, VXMD(i,r,s));
Formula (all,h,CGDS_COMM)(all,r,REG)
VOM(h,r) = VOA(h,r);

Coefficient (all, r, REG)
NETINV(r) # regional NET investment in region r #;
Formula (all, r, REG)
NETINV(r) = sum(k,CGDS_COMM, VOA(k,r)) - VDEP(r);

Coefficient
GLOBINV # global expenditures on net investment #;
Formula
GLOBINV = sum(r,REG, NETINV(r));

Coefficient (all,r,REG)
EK(r) # capital earnings #;
Formula (all,r,REG)
EK(r) = sum(j, ENDWC_COMM, VOA(j,r));

Coefficient
GLOBEK # global capital earnings #;
Formula
GLOBEK = sum{r, REG, EK(r) };

Coefficient (all, r, REG)
YQHHLD(r) # regional household equity income #;
Formula (all, r, REG)
YQHHLD(r) = YQHFIRM(r) + YQHTRUST(r);

Coefficient (all,r,REG)
YQ_FIRM(r) # regional income from capital #;
Formula (all,r,REG)
YQ_FIRM(r) = sum(h,ENDWC_COMM, VOA(h,r)) - VDEP(r);

Coefficient (all,r,REG)
YQ_FHHLDSHR(r) # local capital income share #;
Formula (all, r, REG)
YQ_FHHLDSHR(r) = YQHFIRM(r)/YQ_FIRM(r);

Coefficient (all,r,REG)
WQ_FHHLDSHR(r) # local capital ownership share #;
Formula (all, r, REG)
WQ_FHHLDSHR(r) = YQ_FHHLDSHR(r);

Coefficient (all,r,REG)
WQHFIRM(r) # local ownership of local capital #;
Formula (all,r,REG)
WQHFIRM(r) = WQ_FHHLDSHR(r)*VK(r);

Coefficient
YQ_TRUST # income payments by global fund #;
Formula
YQ_TRUST = sum{r,REG, YQHTRUST(r)};

Coefficient (all,r,REG)
YQ_THHLDSHR(r) # share of region r in income from the global fund #;
Formula (all,r,REG)
YQ_THHLDSHR(r) = YQHTRUST(r)/YQ_TRUST ;

Coefficient (all,r,REG)
WQ_THHLDSHR(r) # share of region r in the global fund #;
Formula (all,r,REG)
WQ_THHLDSHR(r) = YQ_THHLDSHR(r);

Coefficient (all,r,REG)
YQ_FTRUSTSHR(r) # foreign capital income share #;
Formula (all, r, REG)
YQ_FTRUSTSHR(r) = YQTFIRM(r)/YQ_FIRM(r) ;

Coefficient (all,r,REG)
WQ_FTRUSTSHR(r) # foreign capital ownership share #;
Formula (all, r, REG)
WQ_FTRUSTSHR(r) = YQ_FTRUSTSHR(r);

Coefficient (all,r,REG)
WQTFIRM(r) # foreign ownership of local capital #;
Formula (all,r,REG)
WQTFIRM(r) = WQ_FTRUSTSHR(r)*VK(r);

Coefficient
WQTRUST # value of global trust #;
Formula
WQTRUST = sum{r,REG, WQTFIRM(r)};

Coefficient
WQ_TRUST # total regional equity in the global trust #;
Formula
WQ_TRUST = WQTRUST;

Coefficient (all,r,REG)
WQHTRUST(r) # local ownership of foreign property #;
Formula (all,r,REG)
WQHTRUST(r) = WQ_THHLDSHR(r)*WQ_TRUST ;

Coefficient (all,r,REG)
WQHHLD(r) # wealth of the regional household #;
Formula (all,r,REG)
WQHHLD(r) − WQHFIRM(r) + WQHTRUST(r);

Regional Expenditure and Income

Regional income is allocated between private consumption expenditure, government consumption expenditure, and saving.

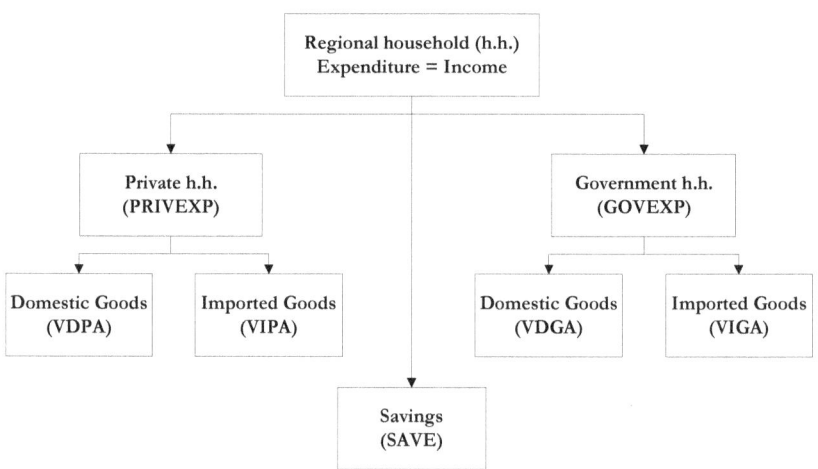

Figure. Expenditure of Regional Household

Note: The coefficients at the ends of branches are Base Data, e.g., VDPA, SAVE.

! government consumption expenditure, GOVEXP !

Coefficient (all,i,TRAD_COMM)(all,r,REG)
VGA(i,r) # government consn expenditure on i in r at agent's prices #;
Formula (all,i,TRAD_COMM)(all,s,REG)
VGA(i,s) = VDGA(i,s) + VIGA(i,s);

Coefficient (all,r,REG)
GOVEXP(r) # government expenditure in region r #;
Formula (all,r,REG)
GOVEXP(r) = sum(i,TRAD_COMM, VGA(i,r));

! private consumption expenditure, PRIVEXP !

Coefficient (all,i,TRAD_COMM)(all,r,REG)
VPA(i,r) # private hhld expenditure on i in r valued at agent's prices #;
Formula (all,i,TRAD_COMM)(all,s,REG)
VPA(i,s) = VDPA(i,s) + VIPA(i,s);

Coefficient (all,r,REG)
PRIVEXP(r) # private consumption expenditure in region r #;
Formula (all,r,REG)
PRIVEXP(r) = sum(i,TRAD_COMM, VPA(i,r));

! aggregate expenditure + saving = aggregate income !

Coefficient (all,r,REG)
INCOME(r) # level of expenditure, which equals NET income in region r #;
Formula (all,r,REG)
INCOME(r) = PRIVEXP(r) + GOVEXP(r) + SAVE(r);

Indirect Tax Receipts

Coefficient (all,i,TRAD_COMM)(all,r,REG)
DGTAX(i,r) # tax on government consumption of domestic good i in region r #;
Formula (all,i,TRAD_COMM)(all,r,REG)
DGTAX(i,r) = VDGA(i,r) - VDGM(i,r);

Coefficient (all,i,TRAD_COMM)(all,r,REG)
IGTAX(i,r) # tax on government consumption of imported good i in region r #;
Formula (all,i,TRAD_COMM)(all,r,REG)
IGTAX(i,r) = VIGA(i,r) - VIGM(i,r);

Coefficient (all,r,REG)
TGC(r) # government consumption tax payments in r #;
Formula (all,r,REG)
TGC(r) = sum(i,TRAD_COMM, DGTAX(i,r) + IGTAX(i,r));

Coefficient (all,i,TRAD_COMM)(all,r,REG)
DPTAX(i,r) # tax on private consumption of domestic good i in region r #;
Formula (all,i,TRAD_COMM)(all,r,REG)
DPTAX(i,r) = VDPA(i,r) - VDPM(i,r);

Coefficient (all,i,TRAD_COMM)(all,r,REG)
IPTAX(i,r) # tax on private consumption of imported good i in region r #;
Formula (all,i,TRAD_COMM)(all,r,REG)
IPTAX(i,r) = VIPA(i,r) - VIPM(i,r);

Coefficient (all,r,REG)
TPC(r) # private consumption tax payments in r #;
Formula (all,r,REG)
TPC(r) = sum(i,TRAD_COMM, DPTAX(i,r) + IPTAX(i,r));

Coefficient (all,i,TRAD_COMM)(all,j,PROD_COMM)(all,r,REG)
DFTAX(i,j,r) # tax on use of domestic intermediate good i by j in r #;
Formula (all,i,TRAD_COMM)(all,j,PROD_COMM)(all,r,REG)
DFTAX(i,j,r) = VDFA(i,j,r) - VDFM(i,j,r);

Coefficient (all,i,TRAD_COMM)(all,j,PROD_COMM)(all,r,REG)
IFTAX(i,j,r) # tax on use of imported intermediate good i by j in r #;
Formula (all,i,TRAD_COMM)(all,j,PROD_COMM)(all,r,REG)
IFTAX(i,j,r) = VIFA(i,j,r) - VIFM(i,j,r);

Coefficient (all,r,REG)
TIU(r) # firms' tax payments on intermediate goods usage in r #;
Formula (all,r,REG)
TIU(r) = sum(i,TRAD_COMM, sum(j,PROD_COMM, DFTAX(i,j,r) + IFTAX(i,j,r)));

Coefficient (all,i,ENDW_COMM)(all,j,PROD_COMM)(all,r,REG)
ETAX(i,j,r) # tax on use of endowment good i by industry j in region r #;
Formula (all,i,ENDW_COMM)(all,j,PROD_COMM)(all,r,REG)
ETAX(i,j,r) = VFA(i,j,r) - VFM(i,j,r);

Coefficient (all,r,REG)
TFU(r) # firms' tax payments on primary factor usage in r #;
Formula (all,r,REG)
TFU(r) = sum(i,ENDW_COMM, sum(j,PROD_COMM, ETAX(i,j,r)));

Coefficient (all,i,NSAV_COMM)(all,r,REG)
PTAX(i,r) # output tax on good i in region r #;
Formula (all,i,NSAV_COMM)(all,r,REG)
PTAX(i,r) = VOM(i,r) - VOA(i,r);

Coefficient (all,r,REG)
TOUT(r) # production tax payments in r #;
Formula (all,r,REG)
TOUT(r) = sum(i,PROD_COMM, PTAX(i,r));

Coefficient (all,i,TRAD_COMM)(all,r,REG)(all,s,REG)
XTAXD(i,r,s) # tax on exports of good i from source r to destination s #;
Formula (all,i,TRAD_COMM)(all,r,REG)(all,s,REG)
XTAXD(i,r,s) = VXWD(i,r,s) - VXMD(i,r,s);

Coefficient (all,r,REG)
TEX(r) # export tax payments in r #;
Formula (all,r,REG)
TEX(r) = sum(i,TRAD_COMM, sum(s,REG, XTAXD(i,r,s)));

Coefficient (all,i,TRAD_COMM)(all,r,REG)(all,s,REG)
MTAX(i,r,s) # tax on imports of good i from source r in destination s #;
Formula (all,i,TRAD_COMM)(all,r,REG)(all,s,REG)
MTAX(i,r,s) = VIMS(i,r,s) - VIWS(i,r,s);

Coefficient (all,r,REG)
TIM(r) # import tax payments in r #;
Formula (all,r,REG)
TIM(r) = sum(i,TRAD_COMM, sum(s,REG, MTAX(i,s,r)));

Miscellaneous Coefficients

! domestic/imported substitution elasticity !

Coefficient (parameter)(all,i,TRAD_COMM)
ESUBD(i)
# region-generic el. of sub. domestic/imported for all agents #;
Read
ESUBD from file GTAPPARM header "ESBD";

! elasticity of cost wrt utility from private consumption !

Coefficient (all,i,TRAD_COMM)(all,r,REG)
CONSHR(i,r) # share of private hhld consumption devoted to good i in r #;
Formula (all,i,TRAD_COMM)(all,r,REG)
CONSHR(i,r) = VPA(i,r) / sum(k,TRAD_COMM, VPA(k,r));

Coefficient (parameter)(all,i,TRAD_COMM)(all,r,REG)
INCPAR(i,r)
# expansion parameter in the CDE minimum expenditure function #;
Read
INCPAR from file GTAPPARM header "INCP";

Coefficient (all,r,REG)
UELASPRIV(r)
# elasticity of cost wrt utility from private consumption #;
Formula (all,r,REG)
UELASPRIV(r) = sum(i,TRAD_COMM, CONSHR(i,r) * INCPAR(i,r));

Appendix B - Modules

This content is adapted from the dynamic GTAP source code.

For more documentation, refer to:
- Hertel, T.W. and M.E. Tsigas "Structure of the Standard GTAP Model", Chapter 2 in T.W. Hertel (editor) *Global Trade Analysis: Modeling and Applications*, Cambridge University Press, 1997.
- Ianchovichina, E.I. (1998) *International Capital Linkages: Theory and Applications in A Dynamic Computable General Equilibrium Model*.
- Ianchovichina, E.I. and R. McDougall, "Theoretical Structure of Dynamic GTAP", *GTAP Technical Paper No. 17*, Dec. 2000

Modules

1. Government Consumption
2. Private Consumption
3. Firms
4. Physical Capital, Global Trust, and Savings
5. International Trade
6. International Transport Services
7. Regional Household
8. Equilibrium Conditions

1. Government Consumption

1-0. Module-Specific Variables
1-1. Demands for Composite Goods
1-2. Composite Tradeables

1-0. Module-Specific Variables

only used in this Government Consumption module

Variable (all,i,TRAD_COMM)(all,r,REG)
pg(i,r) # government consumption price for commodity i in region r #;
Variable (all,i,TRAD_COMM)(all,r,REG)
qg(i,r) # government hhld demand for commodity i in region r #;

1-1. Demands for Composite Goods

Equation GPRICEINDEX
definition of price index for aggregate gov't purchases (HT 40) #   (all,r,REG)   pgov(r) = sum(i,TRAD_COMM, [VGA(i,r) / GOVEXP(r)] * pg(i,r));

Equation GOVDMNDS
government consumption demands for composite commodities (HT 41) #   (all,i,TRAD_COMM)(all,r,REG)   qg(i,r) - pop(r) = ug(r) - [pg(i,r) - pgov(r)];

Equation GOVU
utility from government consumption in r
(all,r,REG)
yg(r) - pop(r) = pgov(r) + ug(r);

1-2. Composite Tradeables

Variable (all,i,TRAD_COMM)(all,r,REG)
tgd(i,r) # tax on domestic i purchased by government hhld in r #;

Variable (all,i,TRAD_COMM)(all,r,REG)
tgm(i,r) # tax on imported i purchased by gov't hhld in r #;

Equation GHHDPRICE
eq'n links domestic market and government consumption prices (HT 19)
(all,i,TRAD_COMM)(all,r,REG)
pgd(i,r) = tgd(i,r) + pm(i,r);

Equation GHHIPRICES
eq'n links domestic market and government consumption prices (HT 22)
(all,i,TRAD_COMM)(all,r,REG)
pgm(i,r) = tgm(i,r) + pim(i,r);

Coefficient (all,i,TRAD_COMM)(all,s,REG)
GMSHR(i,s) # share of imports for gov't hhld at agent's prices #;
Formula (all,i,TRAD_COMM)(all,s,REG)
GMSHR(i,s) = VIGA(i,s) / VGA(i,s);

Equation GCOMPRICE
government consumption price for composite commodities (HT 42)
(all,i,TRAD_COMM)(all,s,REG)
pg(i,s) = GMSHR(i,s) * pgm(i,s) + [1 - GMSHR(i,s)] * pgd(i,s);

Equation GHHLDAGRIMP
government consumption demand for aggregate imports (HT 43) # (all,i,TRAD_COMM)(all,s,REG) qgm(i,s) = qg(i,s) + ESUBD(i) * [pg(i,s) - pgm(i,s)];

Equation GHHLDDOM
government consumption demand for domestic goods (HT 44) # (all,i,TRAD_COMM)(all,s,REG) qgd(i,s) = qg(i,s) + ESUBD(i) * [pg(i,s) - pgd(i,s)];

Equation TGCRATIO
change in ratio of government consumption tax payments to regional income # (all,r,REG) 100.0 * INCOME(r) * del_taxrgc(r) + TGC(r) * y(r) = sum(i,TRAD_COMM, VDGA(i,r) * tgd(i,r) + DGTAX(i,r) * [pm(i,r) + qgd(i,r)]) + sum(i,TRAD_COMM, VIGA(i,r) * tgm(i,r) + IGTAX(i,r) * [pim(i,r) + qgm(i,r)]);

2. Private Consumption Module

2-0. Module-Specific Variables
2-1. Utility from Private Consumption
2-2. Allen Partials, Price and Income Elasticities, Composite Demand
2-3. Composite Tradeables

2-0. Module-Specific Variables
only used in this Private Consumption module

Variable (all,i,TRAD_COMM)(all,r,REG)
pp(i,r) # private consumption price for commodity i in region r #;

Variable (all,i,TRAD_COMM)(all,r,REG)
qp(i,r) # private hhld demand for commodity i in region r #;

2-1. Utility from Private Consumption

Equation PHHLDINDEX
price index for private consumption expenditure #   (all,r,REG)   ppriv(r) = sum(i,TRAD_COMM, CONSHR(i,r) * pp(i,r));

Equation PRIVATEU
computation of utility from private consumption in r (HT 45) #   (all,r,REG)   yp(r) - pop(r) = ppriv(r) + UELASPRIV(r) * up(r);

This equation determines private consumption utility for a representative household in region r, based on the per capita private expenditure function. (HT 45)

Coefficient (all,i,TRAD_COMM)(all,r,REG)
XWCONSHR(i,r)   # expansion-parameter-weighted consumption share #;

Formula (all,i,TRAD_COMM)(all,r,REG)
XWCONSHR(i,r) = CONSHR(i,r) * INCPAR(i,r) / UELASPRIV(r);

Equation UTILELASPRIV
elasticity of expenditure wrt utility from private consumption #   (all,r,REG)   uepriv(r)   = sum(i,TRAD_COMM, XWCONSHR(i,r) * [pp(i,r) + qp(i,r) - yp(r)]);

2-2. Allen Partials, Price and Income Elasticities, Composite Demand

Coefficient (parameter)(all,i,TRAD_COMM)(all,r,REG)
SUBPAR(i,r)   # substitution parameter in CDE minimum expenditure function #;
Read
SUBPAR from file GTAPPARM header "SUBP";

Coefficient (all,i,TRAD_COMM)(all,r,REG)
ALPHA(i,r) # 1 - sub. parameter in the CDE minimum expenditure function #;
Formula (all,i,TRAD_COMM)(all,r,REG)
ALPHA(i,r) = 1 - SUBPAR(i,r);

Coefficient (all,i,TRAD_COMM)(all,k,TRAD_COMM)(all,r,REG)
APE(i,k,r) # Allen partial elst. of sub. between composite i and k in r #;
Formula (all,i,TRAD_COMM)(all,k,TRAD_COMM)(all,r,REG)
APE(i,k,r) = ALPHA(i,r) + ALPHA(k,r) - sum(n,TRAD_COMM, CONSHR(n,r) * ALPHA(n,r));
Formula (all,i,TRAD_COMM)(all,r,REG)
APE(i,i,r) = 2.0 * ALPHA(i,r) - sum(n,TRAD_COMM, CONSHR(n,r) * ALPHA(n,r)) - ALPHA(i,r) / CONSHR(i,r);

Coefficient (all,i,TRAD_COMM)(all,r,REG)
EY(i,r) # income elast. of private hhld demand for i in r (HT F4) #;
Formula (all,i,TRAD_COMM)(all,r,REG)
EY(i,r) = [1.0 / sum(n,TRAD_COMM, CONSHR(n,r) * INCPAR(n,r))] * [INCPAR(i,r) * [1.0 - ALPHA(i,r)] + sum(n,TRAD_COMM, CONSHR(n,r) * INCPAR(n,r) * ALPHA(n,r))] + [ALPHA(i,r) - sum(n,TRAD_COMM, CONSHR(n,r) * ALPHA(n,r))];

Coefficient (all,i,TRAD_COMM)(all,k,TRAD_COMM)(all,r,REG)
EP(i,k,r) # uncomp. elast. of private hhld demand for i wrt price of k in r (HT F5) #;
Formula (all,i,TRAD_COMM)(all,k,TRAD_COMM)(all,r,REG)
EP(i,k,r) = 0;
Formula (all,i,TRAD_COMM)(all,k,TRAD_COMM)(all,r,REG)
EP(i,k,r) = [APE(i,k,r) - EY(i,r)] * CONSHR(k,r);

Equation PRIVDMNDS
private consumption demands for composite commodities (HT 46)
(all,i,TRAD_COMM)(all,r,REG)
qp(i,r) - pop(r) = sum(k,TRAD_COMM, EP(i,k,r) * pp(k,r)) + EY(i,r) * [yp(r) - pop(r)];

Private consumption demands for composite commodities. Demand system is on a per capita basis. Here, yp(r) - pop(r) is % change in per capita income. (HT 46)

2-3. Composite Tradeables

Variable (all,r,REG)
tp(r) # comm.-, source-gen. shift in tax on private cons. #;

The variable tp(r) can be swapped with del_ttaxr(r) in order to generate a tax replacement scenario, whereby taxes remain a constant share of national income.

Variable (all,i,TRAD_COMM)(all,r,REG)
tpd(i,r) # comm.-, source-spec. shift in tax on private cons. of dom. #;

Variable (all,i,TRAD_COMM)(all,r,REG)
atpd(i,r) # power of tax on domestic i purchased by private hhld in r #;

Equation TPDSHIFT
permits uniform consumption tax change # (all,i,TRAD_COMM)(all,r,REG) atpd(i,r) = tpd(i,r) + tp(r);

Equation PHHDPRICE
eq'n links domestic market and private consumption prices (HT 18) # (all,i,TRAD_COMM)(all,r,REG) ppd(i,r) = atpd(i,r) + pm(i,r);

Variable (all,i,TRAD_COMM)(all,r,REG)
tpm(i,r) # comm.-, source-spec. shift in tax on private cons. of imp. #;

Variable (all,i,TRAD_COMM)(all,r,REG)
atpm(i,r) # power of tax on imported i purchased by private hhld in r #;

Equation TPMSHIFT
permits uniform consumption tax change # (all,i,TRAD_COMM)(all,r,REG) atpm(i,r) = tpm(i,r) + tp(r);

Equation PHHIPRICES
eq'n links domestic market and private consumption prices (HT 21) # (all,i,TRAD_COMM)(all,r,REG) ppm(i,r) = atpm(i,r) + pim(i,r);

Equation TPCRATIO
change in ratio of private consumption tax payments to regional income # (all,r,REG) 100.0 * INCOME(r) * del_taxrpc(r) + TPC(r) * y(r) = sum(i,TRAD_COMM, VDPA(i,r) * atpd(i,r) + DPTAX(i,r) * [pm(i,r) + qpd(i,r)]) + sum(i,TRAD_COMM, VIPA(i,r) * atpm(i,r) + IPTAX(i,r) * [pim(i,r) + qpm(i,r)]);

Coefficient (all,i,TRAD_COMM)(all,s,REG)
PMSHR(i,s) # share of imports for priv hhld at agent's prices #;
Formula (all,i,TRAD_COMM)(all,s,REG)
PMSHR(i,s) = VIPA(i,s) / VPA(i,s);

Equation PCOMPRICE
private consumption price for composite commodities (HT 47) # (all,i,TRAD_COMM)(all,s,REG) pp(i,s) = PMSHR(i,s) * ppm(i,s) + [1 - PMSHR(i,s)] * ppd(i,s);

Equation PHHLDDOM
private consumption demand for domestic goods (HT 48) # (all,i,TRAD_COMM)(all,s,REG) qpd(i,s) = qp(i,s) + ESUBD(i) * [pp(i,s) - ppd(i,s)];

Equation PHHLDAGRIMP
private consumption demand for aggregate imports (HT 49) # (all,i,TRAD_COMM)(all,s,REG) qpm(i,s) = qp(i,s) + ESUBD(i) * [pp(i,s) - ppm(i,s)];

3. Firms

We now turn to the behavioral equations for firms. The following picture describes factor demands. The first set of equations describe demands for primary factors. (See table 4 of Hertel and Tsigas.)

Production structure

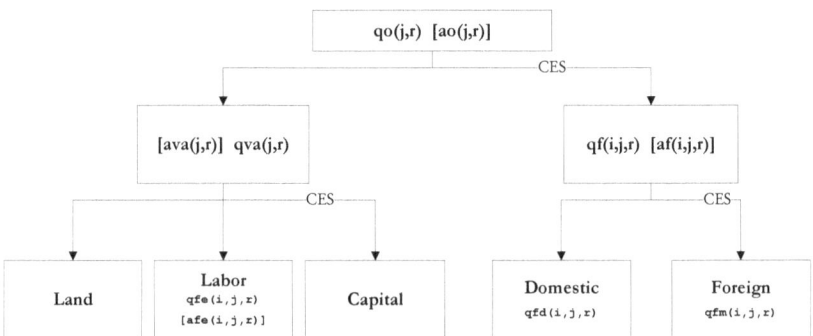

Figure. Production structure

3-0. Module-Specific Variables
3-1. Total Output Nest
3-2. Composite Intermediates Nest
3-3. Value-Added Nest
3-4. Zero Profits Equations

3-0. Module-Specific Variables
only used in this Firms module or the Summary Indices or Welfare appendices

Variable (all,j,PROD_COMM)(all,r,REG)
pva(j,r) # firms' price of value added in industry j of region r #;

Variable (all,j,PROD_COMM)(all,r,REG)
qva(j,r) # value added in industry j of region r #;

Variable (all,i,TRAD_COMM)(all,j,PROD_COMM)(all,r,REG)
pf(i,j,r) # firms' price for commodity i for use by j in r #;

Variable (all,i,TRAD_COMM)(all,j,PROD_COMM)(all,r,REG)
qf(i,j,r) # demand for commodity i for use by j in region r #;

Variable (all,j,PROD_COMM)(all,r,REG)
ao(j,r) # output augmenting technical change in sector j of r #;

Variable (all,i,PROD_COMM)(all,r,REG)
ava(i,r) # value added augmenting tech change in sector i of r #;

Variable (all,i,TRAD_COMM)(all,j,PROD_COMM)(all,r,REG)
af(i,j,r) # composite intermed. input i augmenting tech change by j of r #;

Variable (all,i,ENDW_COMM)(all,j,PROD_COMM)(all,r,REG)
afe(i,j,r) # primary factor i augmenting tech change by j of r #;

Variable (all,i,TRAD_COMM)(all,r,REG)(all,s,REG)
ams(i,r,s) # import i from region r augmenting tech change in region s #;

3-1. Total Output Nest

Variable (all,r,REG)
avareg(r) # Economywide TFP growth rate in region r #;

Variable (all,i,PROD_COMM)
avadiff(i) # Differential rate of TFP growth in sector i #;

Equation TFP
enforces economywide TFP shocks # (all,j,PROD_COMM)(all,r,REG) ava(j,r) = avareg(r) + avadiff(j);

Coefficient (parameter)(all,j,PROD_COMM)
ESUBT(j) # elst. of sub. among composite intermediate inputs in production #;

Read
ESUBT from file GTAPPARM header "ESBT";

Equation VADEMAND
sector demands for primary factor composite # (all,j,PROD_COMM)(all,r,REG) qva(j,r) = -ava(j,r) + qo(j,r) - ao(j,r) - ESUBT(j) * [pva(j,r) - ava(j,r) - ps(j,r) - ao(j,r)];

Sector demands for primary factor composite. This equation differs from HT 35 due to the presence of intermediate input substitution.

Equation INTDEMAND
industry demands for intermediate inputs, including cgds # (all,i,TRAD_COMM)(all,j,PROD_COMM)(all,r,REG) qf(i,j,r) = - af(i,j,r) + qo(j,r) - ao(j,r) - ESUBT(j) * [pf(i,j,r) - af(i,j,r) - ps(j,r) - ao(j,r)];

Industry demands for intermediate inputs, including cgds. This equation differs from HT 36 due to the presence of intermediate input substitution.

3-2. Composite Intermediates Nest

Variable (all,i,TRAD_COMM)(all,j,PROD_COMM)(all,r,REG)
tfd(i,j,r) # tax on domestic i purchased by j in r #;

Equation DMNDDPRICE
eq'n links domestic market and firm prices (HT 20) # (all,i,TRAD_COMM)(all,j,PROD_COMM)(all,r,REG) pfd(i,j,r) = tfd(i,j,r) + pm(i,r);

This equation links domestic market and firm prices. It holds only for domestic goods and it captures the effect of commodity taxation of firms. (HT 20)

Variable (all,i,TRAD_COMM)(all,j,PROD_COMM)(all,r,REG)
tfm(i,j,r) # tax on imported i purchased by j in r #;

Equation DMNDIPRICES
eq'n links domestic market and firm prices (HT 23) # (all,i,TRAD_COMM)(all,j,PROD_COMM)(all,r,REG) pfm(i,j,r) = tfm(i,j,r) + pim(i,r);

This equation links domestic market and firm prices. It holds only for imported goods and it captures the effect of commodity taxation of firms. (HT 23)

Equation TIURATIO
change in ratio of tax payments on intermediate goods to regional income #   (all,r,REG)   100.0 * INCOME(r) * del_taxriu(r) + TIU(r) * y(r)   = sum(i,TRAD_COMM, sum(j,PROD_COMM,   VDFA(i,j,r) * tfd(i,j,r) + DFTAX(i,j,r) * [pm(i,r) + qfd(i,j,r)]))   + sum(i,TRAD_COMM, sum(j,PROD_COMM,   VIFA(i,j,r) * tfm(i,j,r) + IFTAX(i,j,r) * [pim(i,r) + qfm(i,j,r)]));

Coefficient (all,i,TRAD_COMM)(all,j,PROD_COMM)(all,s,REG)
FMSHR(i,j,s) # share of firms' imports in dom. composite, agent's prices #;
Formula (all,i,TRAD_COMM)(all,j,PROD_COMM)(all,s,REG)
FMSHR(i,j,s) = VIFA(i,j,s) / VFA(i,j,s);
Equation ICOMPRICE
industry price for composite commodities (HT 30) #   (all,i,TRAD_COMM)(all,j,PROD_COMM)(all,r,REG)   pf(i,j,r) = FMSHR(i,j,r) * pfm(i,j,r) + [1 - FMSHR(i,j,r)] * pfd(i,j,r);

Equation INDIMP
industry j demands for composite import i (HT 31) #   (all,i,TRAD_COMM)(all,j,PROD_COMM)(all,s,REG)   qfm(i,j,s) = qf(i,j,s) - ESUBD(i) * [pfm(i,j,s) - pf(i,j,s)];

Equation INDDOM
industry j demands for domestic good i (HT 32) #   (all,i,TRAD_COMM)(all,j,PROD_COMM)(all,s,REG)   qfd(i,j,s) = qf(i,j,s) - ESUBD(i) * [pfd(i,j,s) - pf(i,j,s)];

3-3. Value-Added Nest

Variable (all,i,ENDW_COMM)(all,j,PROD_COMM)(all,r,REG)
tf(i,j,r) # tax on primary factor i used by j in region r #;

Equation MPFACTPRICE
eq'n links domestic and firm demand prices (HT 16) # (all,i,ENDWM_COMM)(all,j,PROD_COMM)(all,r,REG) pfe(i,j,r) = tf(i,j,r) + pm(i,r);

Equation SPFACTPRICE
eq'n links domestic and firm demand prices (HT 17) # (all,i,ENDWS_COMM)(all,j,PROD_COMM)(all,r,REG) pfe(i,j,r) = tf(i,j,r) + pmes(i,j,r);

Variable (all,j,PROD_COMM)(all,r,REG)
afesec(j,r) # endowment generic tech change in j of r #;

Variable (all,r,REG)
afereg(r) # Economywide afe shock #;

Coefficient (all,j,PROD_COMM)(all,r,REG)
VVA(j,r) # value added in activity j in region r #;
Formula (all,j,PROD_COMM)(all,r,REG)
VVA(j,r) = sum(i,ENDW_COMM, VFA(i,j,r));

Coefficient (all,i,ENDW_COMM)
SVADEFAULT(i) #zerodivide default for SVA#;
Formula (all,i,ENDW_COMM)
SVADEFAULT(i) = sum(j,PROD_COMM, sum(r,REG, VFA(i,j,r))) / sum(j,PROD_COMM, sum(r,REG, VVA(j,r)));

Coefficient (all,i,ENDW_COMM)(all,j,PROD_COMM)(all,r,REG)
SVA(i,j,r) # share of i in total value added in j in r #;
Formula (all,i,ENDW_COMM)(all,j,PROD_COMM)(all,r,REG: VVA(j,r) <> 0)
SVA(i,j,r) = VFA(i,j,r) / VVA(j,r);
Formula (all,i,ENDW_COMM)(all,j,PROD_COMM)(all,r,REG: VVA(j,r) = 0)
SVA(i,j,r) = SVADEFAULT(i);

Equation VAPRICE
effective price of primary factor composite in each sector/region (HT 33)
(all,j,PROD_COMM)(all,r,REG)
pva(j,r) = sum(k,ENDW_COMM, SVA(k,j,r) * [pfe(k,j,r) - afe(k,j,r)]);

Coefficient (all,i,ENDW_COMM)(all,r,REG)
SVK(i,r) # share of factor i in value added #;
Formula (all,i,ENDW_COMM)(all,r,REG)
SVK(i,r) = EVOA(i,r) / sum(k, ENDW_COMM, EVOA(k,r));

Equation E_AFE
(all,i,ENDWNA_COMM)(all,j,PROD_COMM)(all,r,REG)
afe(i,j,r)
= afereg(r) +((1 - sum(k, ENDWC_COMM, SVK(k,r)))
/ (1 - sum(k, ENDWC_COMM, SVA(k,j,r))))*afesec(j,r);

Equation TFURATIO
change in ratio of tax payments on factor usage to regional income
(all,r,REG)
100.0 * INCOME(r) * del_taxrfu(r) + TFU(r) * y(r)
= sum(i,ENDWM_COMM, sum(j,PROD_COMM,
VFA(i,j,r) * tf(i,j,r) + ETAX(i,j,r) * [pm(i,r) + qfe(i,j,r)]))
+ sum(i,ENDWS_COMM, sum(j,PROD_COMM,
VFA(i,j,r) * tf(i,j,r) + ETAX(i,j,r) * [pmes(i,j,r) + qfe(i,j,r)]));

Coefficient (parameter)(all,j,PROD_COMM)
ESUBVA(j)
# elst. of sub. capital/labor/land, in production of value added in j #;
Read
ESUBVA from file GTAPPARM header "ESBV";

Equation ENDWDEMAND
demands for endowment commodities (HT 34) # (all,i,ENDW_COMM)(all,j,PROD_COMM)(all,r,REG) qfe(i,j,r) = - afe(i,j,r) + qva(j,r) - ESUBVA(j) * [pfe(i,j,r) - afe(i,j,r) - pva(j,r)];

3-4. Zero Profits Equations

Equation OUTPUTPRICES
eq'n links pre- and post-tax supply prices for all industries (HT 15) # (all,i,PROD_COMM)(all,r,REG) ps(i,r) = to(i,r) + pm(i,r);

This equation links pre- and post-tax supply prices for all industries. This captures the effect of output taxes. TO(i,r) < 1 in the case of a tax. (HT 15)

Equation TOUTRATIO
change in ratio of output tax payments to regional income # (all,r,REG) 100.0 * INCOME(r) * del_taxrout(r) + TOUT(r) * y(r) = sum(i,PROD_COMM, VOA(i,r) * [-to(i,r)] + PTAX(i,r) * [pm(i,r) + qo(i,r)]);

Variable (all,j,PROD_COMM)(all,r,REG)
profitslack(j,r) # slack variable in the zero profit equation #;

This is exogenous, unless the user wishes to specify output in a given region exogenously.

Coefficient (all,i,DEMD_COMM)(all,j,PROD_COMM)(all,r,REG)
STC(i,j,r) # share of i in total costs of j in r #;
Formula (all,i,DEMD_COMM)(all,j,PROD_COMM)(all,r,REG)
STC(i,j,r) = VFA(i,j,r) / sum(k,DEMD_COMM, VFA(k,j,r));

Equation ZEROPROFITS
industry zero pure profits condition (HT 6) # (all,j,PROD_COMM)(all,r,REG) ps(j,r) + ao(j,r) = sum(i,ENDW_COMM, STC(i,j,r) * [pfe(i,j,r) - afe(i,j,r) - ava(j,r)]) + sum(i,TRAD_COMM, STC(i,j,r) * [pf(i,j,r) - af(i,j,r)]) + profitslack(j,r);

Industry zero pure profits condition (HT 6). This condition permits us to determine the endogenous output level for each of the non-endowment sectors, excepting when profitslack is itself endogenous. The level of activity in the endowment sectors is exogenously determined.

4. Physical Capital, Global Trust, and Savings

Dynamics extension: investment theory

These equations determine investment by region, `qcgds'. They also determine descriptive variables representing regional rates of return (`DROR') and the world average rate of return (`DRORW').

That's right, the rate of return variable is now merely descriptive. Investment responds to the expected not the actual rate. The actual rates does indirectly affect behavior, via an expectations adjustment process. But as it happens, it appears there as a coefficient not as a variable.

Another change: the rate of return variable is now an absolute change variable. This is appropriate since the (net) rate of return may be either positive or negative. We also include variables pertaining to (expected and target) gross rates of return: these are percentage change variables, because the gross rate of return is always positive; because however high the stock of capital relative to the demand, there is always some strictly positive capital rental low enough to
clear the market.

Some change in nomenclature: we use `qk' for capital stock, not `kb', because we don't now have separate variables for beginning- and end-of-period capital stocks, because we don't have a period, just a point in time. We use `qk' not 'k' because `k' appears elsewhere as an index. Also, we use `VK' for the capital stock coefficient, not `VKB'.

4-0. Module-Specific Variables
4-1. Equations of Notational Convenience
4-2. Rate of Return Equations
4-3. Required Rate of Growth in Rate of Return
4-4. Investment
4-5. Capital
4-6. Global Trust
4-7. Saving

4-0. Module-Specific Variables

only used in this Physical Capital, Global Trust, and Savings module

Variable (change)(all,r,REG)
DKHAT(r) # normal rate of growth in capital #;

Variable (all, r, REG)
rental(r) # rental rate on capital = ps("capital",r) #;

Variable (all,r,REG)
rorga(r) # actual rate of return #;

Variable (change)(all,r,REG)
erg_rorg(r) # expected rate of growth in gross rate of return #;

Variable (all,r,REG)
rorgt(r) # target gross rate of return #;

Variable (all,r,REG)
rorge(r) # expected gross rate of return #;

Coefficient (parameter)(all,r,REG)
RORGFLEX(r) # flexibility of gross rate of return #;
Read
RORGFLEX from file GTAPPARMK header "RGFX";

Agents expect each 1% expansion in the capital stock to reduce the gross rate of return by RORGFLEX %

Coefficient (all, r, REG)
REGINV(r) # regional GROSS investment in r (value of "cgds" output) #;
Formula (all, r, REG)
REGINV(r) = sum(k,CGDS_COMM, VOA(k,r));

Variable (levels)(all, r, REG)
QREGINV(r) # real GROSS investment in r (qty of "cgds" output) #;
Formula (initial)(all, r, REG)
QREGINV(r) = sum(k,CGDS_COMM, VOA(k,r));
Equation E_QREGINV(all, r, REG)
p_QREGINV(r) = qcgds(r) ;

Variable (levels)(all, r, REG)
QREGINV_D(r) # real GROSS investment in r (qty determined by dynamics) #;
Formula (initial)(all, r, REG)
QREGINV_D(r) = sum(k,CGDS_COMM, VOA(k,r));
Equation E_QREGINV_D(all, r, REG)
p_QREGINV_D(r) = qcgds_d(r) ;

Coefficient (parameter)
MINIKRAT # Minimum ratio of Investment to Capital stock #;
Read
MINIKRAT from file GTAPPARMK header "MIKR";

Variable (levels)(all,r,REG)
QMININV(r) # Minimum GROSS investment allowed in r #;
Formula (initial) (all,r,REG)
QMININV(r) = MINIKRAT * VK(r) ;
Equation E_MININV(all, r, REG)
p_QMININV(r) = qk(r) ;

Complementarity (variable = QREGINV, lower_bound = QMININV)
! Ensures QREGINV = QREGINV_D when QREGINV_D > QMININV else QREGINV = QMININV ! C_QREGINV (all,r,REG) QREGINV(r) - QREGINV_D(r) ;

Coefficient (all,r,REG)
KHAT(r) # price-neutral rate of growth in the capital stock #;
Update (change) (all,r,REG)
KHAT(r) = (1/100) * DKHAT(r);
Read
KHAT from file GTAPDATA header "KHAT";

the rate of growth in the capital stock which is, agents believe, consistent with a constant rate of return on capital

4-1. Equations of Notational Convenience

Equation KAPSVCES
eq'n defines a variable for capital services (HT 52) # (all,r,REG) qk(r) = sum(h,ENDWC_COMM, [VOA(h,r) / sum(k,ENDWC_COMM, VOA(k,r))] * qo(h,r));

This equation defines the relationship between beginning capital stock and the endowment of capital. (There is really only one capital services item.) (HT 52)

Equation KAPRENTAL
eq'n defines a variable for capital rental rate (HT 53) # (all,r,REG) rental(r) = sum(h,ENDWC_COMM, [VOA(h,r) / sum(k,ENDWC_COMM, VOA(k,r))] * ps(h,r));

Equation CAPGOODS
eq'n defines a variable for gross investment (HT 54) # (all,r,REG) qcgds(r) = sum(h,CGDS_COMM, [VOA(h,r) / REGINV(r)] * qo(h,r));

This equation defines a variable for gross investment, for convenience. There is really only one capital goods item.

Equation PRCGOODS
eq'n defines the price of cgds (HT 55) # (all,r,REG) pcgds(r) = sum(h,CGDS_COMM, [VOA(h,r) / REGINV(r)] * ps(h,r));

Variable (all, r, REG)
sqk(r) # Arbitrary region-specific shock to capital stock #;

Variable
sqkworld # Arbitrary region-generic shock to capital stock #;

Equation KBEGINNING
associates change in cap. services w/ change in cap. stock (HT 56)
(all,r,REG)
VK(r) * qk(r) = 100 * NETINV(r) * time + VK(r) * [sqk(r) + sqkworld];

This equation associates any change in capital services during the period with a change in capital stock. Full capacity utilization is assumed.

4-2. Rate of Return Equations

Coefficient (parameter)(all,r,REG)
LAMBRORGE(r) # coefficient of adjustment for expected rate of return #;
Read
LAMBRORGE from file GTAPPARMK header "LRGE";

controls the rate at which agents adjust expectations of rates of return in response to differences between expected and actual rates. If the actual rate exceeds the expected rate by 1%, agents adjust the expected rate by LAMBRORGE% per period.

Coefficient (ge 0)(all,r,REG)
RORGEXP(r) # expected gross rate of return #;
Update (all,r,REG)
RORGEXP(r) = rorge(r);
Read
RORGEXP from file GTAPDATA header "RRGE";

Coefficient (ge 0)(all,r,REG)
RORGROSS(r) # gross rate of return #;
Formula (all,r,REG)
RORGROSS(r) = EK(r) / VK(r);

Coefficient (ge 0, le 10)(all,r,REG)
ARGLOG(r) # argument to loge #;
Formula (all,r,REG)
ARGLOG(r) = RORGEXP(r)/RORGROSS(r);

Coefficient (all,r,REG)
ERRRORG(r) # measure of error in rate of return #;
Formula (all,r,REG)
ERRRORG(r) = loge(ARGLOG(r));

Variable (change)(all,r,REG)
erg_rorg_d(r) # expected rate of growth in gross rate of return #;
Variable (all,r,REG)
rorge_d(r) # expected gross rate of return #;

Variable (all,r,REG)
srorge(r) # expected gross rate of return #;
Equation EXPECTED_ROR
rule for expected gross rate of return #   (all,r,REG)   rorge_d(r)   = - RORGFLEX(r) * [qk(r) - 100.0 * KHAT(r) * time]   - 100.0 * LAMBRORGE(r) * ERRRORG(r) * time   + srorge(r);

Variable (change)(all,r,REG)
DRORT(r) # target rate of return #;
Variable (change)
SDRORTW # world-wide shift in target rate of return #;
Variable (change)(all,r,REG)
SDRORT(r)# region-specific shift in target rate of return #;
Equation NET_ROR
equilibrium condition for rate of return #   (all,r,REG)   DRORT(r) = SDRORTW + SDRORT(r);

Coefficient (ge 0)(all,r,REG)
RORGTARG(r) # target gross rate of return #;
Update (all,r,REG)
RORGTARG(r) = rorgt(r) ;
Read
RORGTARG from file GTAPDATA header "RRGT";

Equation GROSS_ROR
identity for target gross rate of return #   (all,r,REG)   RORGTARG(r) * rorgt(r) = DRORT(r);

Equation RATERETURNP
identity for rate of return
(all,r,REG)
rorga(r) = rental(r) - pcgds(r);

Variable (change)(all,r,REG)
DROR(r) # rate of return #;

net or gross

Equation RATERETURN
identity for net rate of return
(all,r,REG)
DROR(r) = RORGROSS(r) * [rental(r) - pcgds(r)] ;

Replaces (HT 57)

Coefficient
GLOBVK # world stock of capital #;
Formula
GLOBVK = sum{r,REG, VK(r)};

Variable (change)
DRORW # world-wide average rate of return #;

net or gross

Coefficient (all,r,REG)
RORNET(r) # net rate of return #;
Formula (all,r,REG)
RORNET(r) = [EK(r) - VDEP(r)] / VK(r);

Coefficient
RORNWORLD # world average net rate of return #;
Formula
RORNWORLD = [GLOBEK - sum{r,REG, VDEP(r)}] / GLOBVK ;

Equation WORLDAVROR
identity for world average rate of return #   GLOBVK * DRORW   = sum{r,REG, VK(r) * DROR(r) + [RORNET(r) - RORNWORLD] * VK(r) * [pcgds(r) + qk(r)]};

Variable (change)
SDRORW # world-wide shift in rate of return #;
Variable (change)(all,r,REG)
SDROR(r) # region-specific shift in rate of return #;
Equation NET_ROR_EXTRA
equilibrium condition for rate of return #   (all,r,REG)   DROR(r) = SDRORW + SDROR(r);

Variable (all, r, REG)
rorwqht(r) # rate of return to foreign equity owned by a region #;
Equation RRWRTEQY
equation defines rate of return to foreign equity owned by a region #   (all,r,REG)   rorwqht(r) = yqt - wq_t;

Variable (all, r, REG)
rorwqtf(r) # rate of return to equity owned by foreigners in a region #;
Equation RRWTFEQY
eqn defines rate of return to equity owned by foreigners in a reg #   (all,r,REG)   rorwqtf(r) = yq_f(r) - wq_f(r);

4-3. Required Rate of Growth in Rate of Return

Coefficient (all,r,REG)
IKRATIO(r) # investment:capital stock ratio #;
Formula (all,r,REG)
IKRATIO(r) = REGINV(r) / VK(r);

Equation EGROWTH_D_ROR
rule for expected rate of growth in rate of return #   (all,r,REG)   erg_rorg_d(r)   = - RORGFLEX(r) * IKRATIO(r) * [qcgds_d(r) - qk(r)]   + RORGFLEX(r) * DKHAT(r);

Equation EGROWTH_ROR
rule for expected rate of growth in rate of return #   (all,r,REG)   erg_rorg(r)   = - RORGFLEX(r) * IKRATIO(r) * [qcgds(r) - qk(r)]   + RORGFLEX(r) * DKHAT(r);

4-4. Investment

Coefficient (parameter)(all,r,REG)
LAMBRORG(r) # coefficient of adjustment for rate of return #;
Read
LAMBRORG from file GTAPPARMK header "LRRG";

controls the rate at which agents aim to adjust rates of return in response to differences between expected and target rates. If the target rate exceeds the expected rate by 1%, agents aim to adjust the rate by LAMBRORG% per period.

Equation INVESTMENT
rule for investment
(all,r,REG)
erg_rorg_d(r) = LAMBRORG(r) * [rorgt(r) - rorge_d(r)];

Equation E_RORGE
rule for investment
(all,r,REG)
erg_rorg(r) = LAMBRORG(r) * [rorgt(r) - rorge(r)];

The global Trust allocates investment across regions in a way which will, it expects, bring the rate of return gradually into line with the target rate.

Variable (all,r,REG)
sqcgdsreg(r) # arbitrary region specific shock to investment #;
Variable
sqcgdsworld # arbitrary region generic shock to investment #;
Equation GDI
region specific determination of investment #   (all,r,REG)   qcgds(r) = sqcgdsreg(r) + sqcgdsworld;

Variable
globalcgds # Global supply of capital goods for NET investment #;
Equation GLOBALINV
This equation computes: the change in global investment #   globalcgds   = sum(r,REG, {REGINV(r)/GLOBINV} * qcgds(r)   - {VDEP(r)/GLOBINV} * qk(r));

4-5. Capital

Coefficient (parameter)(all,r,REG)
LAMBKHAT(r) # coefficient of adjustment for the normal rate of growth #;
Read
LAMBKHAT from file GTAPPARMK header "LKHT";

Variable (change) (all,r,REG)
SDKHAT(r) # normal rate of growth in capital #;
Equation KHATGROWTH
rule for normal rate of growth in capital #   (all,r,REG)   DKHAT(r)   = LAMBKHAT(r) * [qk(r) + (1 / RORGFLEX(r)) * rorga(r)   - 100 * KHAT(r) * time] + SDKHAT(r);

Variable (all,r,REG)
swq_f(r) # shift in wealth of region r #;
Equation REGEQYLCL
This equation determines the change in VK(r) #   (all, r, REG)   wq_f(r) = qk(r) + pcgds(r) + swq_f(r);

Coefficient (all,r,REG)
WQ_FIRM(r) # equity located in region r #;
Formula (all,r,REG)
WQ_FIRM(r) = WQHFIRM(r) + WQTFIRM(r);

Equation EQYHOLDFNDLCL
eq'n determines shift variable for value of domestic capital #   (all,r,REG)   WQ_FIRM(r) * wq_f(r)   = WQHFIRM(r) * wqhf(r) + WQTFIRM(r) * wqtf(r);

4-6. Global Trust

Coefficient
YQTRUST # income of the global trust #;

Formula
YQTRUST = sum(r, REG, YQTFIRM(r));

Equation PKWRLD
eq'n determines the change in the price of equity in the global trust #   WQTRUST * pqtrust = sum(r,REG, WQTFIRM(r) * pcgds(r));

Equation INCFNDLCLEQY
eq'n determines the income of the global fund from equity in region r #   (all, r, REG)   yqtf(r) = yq_f(r) + wqtf(r) - wq_f(r);

Equation TOTGFNDASSETS
eq'n determines the change in total assets of the global fund #   WQTRUST * wqt = sum(s, REG, WQTFIRM(s) * wqtf(s));

Equation TOTGFNDPROP

# eq'n determines the change in total proprietorship in the global trust # 	
 wq_t = sum(s, REG, [WQHTRUST(s)/WQ_TRUST] * wqht(s));

Equation INCFNDEQY

# eq'n determines the change in the income of the global trust # 	
 yqt = sum(r, REG, [YQTFIRM(r) / YQTRUST] * yqtf(r));

4-7. Saving

Equation PRISAV
eq'n for a price index for the aggregate global cgds composite #   (all,r,REG)   psave(r) = (WQHFIRM(r)/WQHHLD(r)) * pcgds(r)   + (WQHTRUST(r)/WQHHLD(r)) * pqtrust;

Variable
psavewld # price of capital goods supplied to savers #;
Equation PSAVEWORLD

# eq'n generates a price index for the aggregate global cgds composite # 	
 psavewld = sum(r,REG, [NETINV(r) / GLOBINV] * pcgds(r));

5. International Trade

5-1. Export Prices
5-2. Demand for Imports

5-1. Export Prices

Variable (all,i,TRAD_COMM)(all,r,REG)
tx(i,r) # dest.-gen. change in subsidy on exports of i from r #;

Variable (all,i,TRAD_COMM)(all,r,REG)(all,s,REG)
txs(i,r,s) # dest.-spec. change in subsidy on exports of i from r to s #;

The variable txs captures changes in the power of bilateral export taxes. However, the presence of a destination-generic export subsidy shift (tx) also permits the user to swap a single export tax shock with another target variable. It is most naturally swapped with the variable qo to insulate domestic producers from the world market.

Equation EXPRICES
eq'n links agent's and world prices (HT 27) # (all,i,TRAD_COMM)(all,r,REG)(all,s,REG) pfob(i,r,s) = pm(i,r) - tx(i,r) - txs(i,r,s);

Equation TEXPRATIO
change in ratio of export tax payments to regional income # (all,r,REG) 100.0 * INCOME(r) * del_taxrexp(r) + TEX(r) * y(r) = sum(i,TRAD_COMM, sum(s,REG, VXMD(i,r,s) * [-tx(i,r) - txs(i,r,s)] + XTAXD(i,r,s) * [pfob(i,r,s) + qxs(i,r,s)]));

5-2. Demand for Imports

Composite Imports Nest: Table 3 of Hertel and Tsigas

Variable (all,i,TRAD_COMM)(all,s,REG)
tm(i,s) # source-gen. change in tax on imports of i into s #;

Variable (all,i,TRAD_COMM)(all,r,REG)(all,s,REG)
tms(i,r,s) # source-spec. change in tax on imports of i from r into s #;

The variable tms captures changes in the power of bilateral import taxes. However, the presence of a source-generic import tariff shift (tm) also permits the user to swap a single import tariff shock with another target variable. In particular, to insulate domestic producers from import price changes, it may be swapped with the relative price variable pr -- see below.

Equation MKTPRICES
eq'n links domestic and world prices (HT 24)
(all,i,TRAD_COMM)(all,r,REG)(all,s,REG)
pms(i,r,s) = tm(i,s) + tms(i,r,s) + pcif(i,r,s);

Coefficient (all,i,TRAD_COMM)(all,r,REG)(all,s,REG)
MSHRS(i,r,s) # share of imports from r in import bill of s at mkt prices #;
Formula (all,i,TRAD_COMM)(all,r,REG)(all,s,REG)
MSHRS(i,r,s) = VIMS(i,r,s) / sum(k,REG, VIMS(i,k,s));
Equation DPRICEIMP
price for aggregate imports (HT 28)
(all,i,TRAD_COMM)(all,s,REG)
pim(i,s) = sum(k,REG, MSHRS(i,k,s) * [pms(i,k,s) - ams(i,k,s)]);

Variable (orig_level=1.0)(all,i,TRAD_COMM)(all,r,REG)
pr(i,r) # ratio of domestic to imported prices in r #;
Equation PRICETGT
eq'n defines target price ratio to be attained via the variable levy (HT 25) # (all,i,TRAD_COMM)(all,s,REG) pr(i,s) = pm(i,s) - pim(i,s);

This equation defines the target price ratio to be attained via the variable levy. This price ratio is the ratio of domestic to average imported goods' price. Note that the way this price ratio is defined, it includes intraregional imports as well. In most applications, regions will present groups of individual countries. However, in the case of the EU, this is problematic, since recent versions of the database have incorporated intra-EU trade flows. Therefore, when aggregated to the EU level, the composite import price includes both intra-EU and outside imports. So some modification is needed to handle the EU case.

Coefficient (parameter)(all,i,TRAD_COMM)
ESUBM(i)
# region-generic el. of sub. among imports of i in Armington structure #;
Read
ESUBM from file GTAPPARM header "ESBM";

Equation IMPORTDEMAND
regional demand for disaggregated imported commodities by source (HT 29) # (all,i,TRAD_COMM)(all,r,REG)(all,s,REG) qxs(i,r,s) = -ams(i,r,s) + qim(i,s) - ESUBM(i) * [pms(i,r,s) - ams(i,r,s) - pim(i,s)];

Equation TIMPRATIO
change in ratio of import tax payments to regional income #   (all,r,REG)   100.0 * INCOME(r) * del_taxrimp(r) + TIM(r) * y(r)   = sum(i,TRAD_COMM, sum(s,REG,   VIMS(i,s,r) * [tm(i,r) + tms(i,s,r)]   + MTAX(i,s,r) * [pcif(i,s,r) + qxs(i,s,r)]));

6. International Transport Services

6-0. Module-Specific Variables and Coefficients
6-1. Demand for Global Transport Services
6-2. Supply of Transport Services

6-0. Module-Specific Variables and Coefficients

only used in this International Transport Services module

Variable
(all,m,MARG_COMM)(all,i,TRAD_COMM)(all,r,REG)(all,s,REG)
qtmfsd(m,i,r,s) # international usage margin m on i from r to s #;

International margin usage, by Margin, Freight, Source, and Destination, i.e., the percent change in usage of m in transport of i from r to s.

Variable
(all,m,MARG_COMM)(all,i,TRAD_COMM)(all,r,REG)(all,s,REG)
atmfsd(m,i,r,s) # tech change in m's shipping of i from region r to s #;

Technical progress in shipping by Margin, Freight, Source, and Destination. This is endogenous and driven by the following mode-, product-, source-, and destination-specific determinants.

Variable (all,m,TRAD_COMM)
atm(m) # tech change in mode m, worldwide #;

Variable (all,i,TRAD_COMM)
atf(i) # tech change shipping of i, worldwide #;

Variable (all,r,REG)
ats(r) # tech change shipping from region r #;

Variable (all,s,REG)

atd(s) # tech change shipping to s #;

Variable (all,m,MARG_COMM)(all,i,TRAD_COMM)(all,r,REG)(all,s,REG)
atall(m,i,r,s) # tech change in m's shipping of i from region r to s #;

Variable (all,i,TRAD_COMM)(all,r,REG)(all,s,REG)
ptrans(i,r,s) # cost index for international transport of i from r to s #;

average cost index for margin services used in getting i from r to s

Variable (all,m,MARG_COMM)
qtm(m) # global margin usage #;

Variable (all,m,MARG_COMM)
pt(m) # price of composite margins services, type #;

price index for commodity m in margin services usage

Coefficient (ge 0) (all,m,MARG_COMM)(all,i,TRAD_COMM)(all,r,REG)(all,s,REG) VTMFSD(m,i,r,s) # int'l margin usage, by margin, freight, source, and destination #;
Update (all,m,MARG_COMM)(all,i,TRAD_COMM)(all,r,REG)(all,s,REG) VTMFSD(m,i,r,s) = pt(m) * qtmfsd(m,i,r,s);
Read
VTMFSD from file GTAPDATA header "VTWR";
Coefficient (all,i,TRAD_COMM)(all,r,REG)(all,s,REG)
VTFSD(i,r,s) # aggregate value of svces in the shipment of i from r to s #;
Formula (all,i,TRAD_COMM)(all,r,REG)(all,s,REG)
VTFSD(i,r,s) = sum(m,MARG_COMM, VTMFSD(m,i,r,s));

In a balanced data base, (all,i,TRAD_COMM)(all,r,REG)(all,s,REG), VIWS(i,r,s) = VXWD(i,r,s) + VTFSD(i,r,s).

Coefficient (all,m,MARG_COMM)
VTMUSE(m) # international margin services usage, by type #;
Formula (all,m,MARG_COMM)
VTMUSE(m) = sum(i,TRAD_COMM, sum(r,REG, sum(s,REG, VTMFSD(m,i,r,s))));

Coefficient (all,m,MARG_COMM)
VTMPROV(m) # international margin services provision #;
Formula (all,m,MARG_COMM)
VTMPROV(m) = sum(r,REG, VST(m,r));

In a balanced data base, VTMPROV = VTMUSE.

Coefficient (all,r,REG)
VTRPROV(r) # international margin supply, by region #;
Formula (all,r,REG)
VTRPROV(r) = sum(m,MARG_COMM, VST(m,r));

Coefficient
VT # international margin supply #;
Formula
VT = sum(m,MARG_COMM, sum(r,REG, VST(m,r)));

6-1. Demand for Global Transport Services

Equation QTRANS_MFSD
bilateral demand for transport services #   (all,m,MARG_COMM)(all,i,TRAD_COMM)(all,r,REG)(all,s,REG)   qtmfsd(m,i,r,s) = qxs(i,r,s) - atmfsd(m,i,r,s);

This equation computes the bilateral demand for international transportation services. It reflects the fact that the demand for services along any particular route is proportional to the quantity of merchandise shipped [i.e., QXS(i,r,s)]. It is here that we introduce the potential for input-augmenting tech change, atmfsd(m,i,r,s), which is commodity- and route-specific. Thus, in the levels: ATMFSD(m,i,r,s) * QTMFSD(m,i,r,s) = QXS(i,r,s) where QTMFSD is the amount of composite margins services m used along this route. Technological improvements are reflected by atmfsd(i,r,s) > 0, and these reduce the margins services required for this i,r,s triplet. Tech. change also dampens the cost of shipping, thereby lowering the CIF price implied by a given FOB value (see 6-2).

Coefficient (all,m,MARG_COMM)(all,i,TRAD_COMM)(all,r,REG)(all,s,REG) VTMUSESHR(m,i,r,s) # share of i,r,s usage in global demand for m #;
Formula (all,m,MARG_COMM)(all,i,TRAD_COMM)(all,r,REG)(all,s,REG) VTMUSESHR(m,i,r,s) = VTFSD(i,r,s) / VT;
Formula (all,m,MARG_COMM: VTMUSE(m) <> 0.0)(all,i,TRAD_COMM)(all,r,REG)(all,s,REG) VTMUSESHR(m,i,r,s) = VTMFSD(m,i,r,s) / VTMUSE(m);

Equation TRANS_DEMAND
global demand for margin m # (all,m,MARG_COMM) qtm(m) = sum(i,TRAD_COMM, sum(r,REG, sum(s,REG, VTMUSESHR(m,i,r,s) * qtmfsd(m,i,r,s))));

6-2. Supply of Transport Services

Coefficient (all,m,MARG_COMM)(all,r,REG)
VTSUPPSHR(m,r) # share of region r in global supply of margin m #;
Formula (all,m,MARG_COMM)(all,r,REG)
VTSUPPSHR(m,r) = VTRPROV(r) / VT;
Formula (all,m,MARG_COMM: VTMPROV(m) <> 0.0)(all,r,REG)
VTSUPPSHR(m,r) = VST(m,r) / VTMPROV(m);

Equation PTRANSPORT
generate price index for composite transportation services # (all,m,MARG_COMM) pt(m) = sum(r,REG, VTSUPPSHR(m,r) * pm(m,r));

This equation generates a price index for transportation services based on zero profits. NOTE:
(1) Sales to international transportation are not subject to export tax. This is why we base the costs to the transport sector on market prices of the goods sold to international transportation.
(2) We assume that the supply shares for margin services are uniform across freight, source of freight, and destination. (cf. HT#7)

Coefficient
VTUSE # international margin services usage #;
Formula
VTUSE = sum(m,MARG_COMM, sum(i,TRAD_COMM, sum(r,REG, sum(s,REG, VTMFSD(m,i,r,s)))));

Coefficient
(all,m,MARG_COMM)(all,i,TRAD_COMM)(all,r,REG)(all,s,REG) VTFSD_MSH(m,i,r,s) # share of margin m in cost of getting i from r to s #;
Formula (all,m,MARG_COMM)(all,i,TRAD_COMM)(all,r,REG) (all,s,REG: VTFSD(i,r,s) > 0.0) VTFSD_MSH(m,i,r,s) = VTMFSD(m,i,r,s) / VTFSD(i,r,s);
Formula (all,m,MARG_COMM)(all,i,TRAD_COMM)(all,r,REG) (all,s,REG: VTFSD(i,r,s) = 0.0) VTFSD_MSH(m,i,r,s) = VTMUSE(m) / VTUSE;

Equation TRANSCOSTINDEX
generates flow-specific modal average cost of transport index (cf. HT7) # (all,i,TRAD_COMM)(all,r,REG)(all,s,REG) ptrans(i,r,s) = sum(m,MARG_COMM, VTFSD_MSH(m,i,r,s) * [pt(m) - atmfsd(m,i,r,s)]);

Equation TRANSTECHANGE
generates flow-specific average rate of technical change # (all,m,MARG_COMM)(all,i,TRAD_COMM)(all,r,REG)(all,s,REG) atmfsd(m,i,r,s) = atm(m) + atf(i) + ats(r) + atd(s) + atall(m,i,r,s);

Equation TRANSVCES
generate demand for regional supply of global transportation service (HT 61) # (all,m,MARG_COMM)(all,r,REG) qst(m,r) = qtm(m) + [pt(m) - pm(m,r)];

This equation generates the international transport sector's derived demand for regional supplies of transportation services. It reflects a unitary elasticity of substitution between transportation services inputs from different regions.

Coefficient (all,i,TRAD_COMM)(all,r,REG)(all,s,REG)
VIWSCOST(i,r,s) # value of imports calculated as total cost of imports #;
Formula (all,i,TRAD_COMM)(all,r,REG)(all,s,REG)
VIWSCOST(i,r,s) = VXWD(i,r,s) + VTFSD(i,r,s);

Coefficient (all,i,TRAD_COMM)(all,r,REG)(all,s,REG)
FOBSHR(i,r,s) # FOB share in VIW #;
Formula (all,i,TRAD_COMM)(all,r,REG)(all,s,REG)
FOBSHR(i,r,s) = VXWD(i,r,s) / VIWSCOST(i,r,s);

Coefficient (all,i,TRAD_COMM)(all,r,REG)(all,s,REG)
TRNSHR(i,r,s) # transport share in VIW #;
Formula (all,i,TRAD_COMM)(all,r,REG)(all,s,REG)
TRNSHR(i,r,s) = VTFSD(i,r,s) / VIWSCOST(i,r,s);

Equation FOBCIF

# eq'n links FOB and CIF prices for good i shipped from region r to s (HT 26') # 	
(all,i,TRAD_COMM)(all,r,REG)(all,s,REG)
pcif(i,r,s)
  = FOBSHR(i,r,s) * pfob(i,r,s)
  + TRNSHR(i,r,s) * ptrans(i,r,s);

This equation links export and import prices for each commodity/route triplet. Note that technical change is embodied in ptrans(i,r,s) which is now a cost index, as opposed to (HT 26') where it represented the price of
margins services.

7. Regional Household

7-0. Module-Specific Coefficients
7-1. Supply of Endowments by the Regional Household
7-2. Regional Wealth and Equity Income
7-3. Computation of Regional Income
7-4. Regional Household Demand System
7-5. Aggregate Utility

7-0. Module-Specific Coefficients

only used in this Regional Household module

Coefficient (all,r,REG)
XSHRPRIV(r) # private expenditure share in regional income #;
Formula (all,r,REG)
XSHRPRIV(r) = PRIVEXP(r) / INCOME(r);

Coefficient (all,r,REG)
XSHRGOV(r) # government expenditure share in regional income #;
Formula (all,r,REG)
XSHRGOV(r) = GOVEXP(r) / INCOME(r);

Coefficient (all,r,REG)
XSHRSAVE(r) # saving share in regional income #;
Formula (all,r,REG)
XSHRSAVE(r) = SAVE(r) / INCOME(r);

Variable (all,r,REG)
uelas(r) # elasticity of cost of utility wrt utility #;

Variable (all,r,REG)
dppriv(r) # private consumption distribution parameter #;

Variable (all,r,REG)
dpgov(r) # government consumption distribution parameter #;

Variable (all,r,REG)
dpsave(r) # saving distribution parameter #;

Variable (all,r,REG)
yqh(r) # regional household equity income #;

7-1. Supply of Endowments by the Regional Household

Equation FACTORINCPRICES
eq'n links pre- and post-tax endowment supply prices (HT 15) #   (all,i,ENDW_COMM)(all,r,REG)   ps(i,r) = to(i,r) + pm(i,r);

Coefficient (all,r,REG)
TINC(r) # income tax payments in r #;
Formula (all,r,REG)
TINC(r) = sum(i,ENDW_COMM, PTAX(i,r));

Equation TINCRATIO
change in ratio of income tax payments to regional income #   (all,r,REG)   100.0 * INCOME(r) * del_taxrinc(r) + TINC(r) * y(r)   = sum(i,ENDW_COMM,   VOA(i,r) * [-to(i,r)] + PTAX(i,r) * [pm(i,r) + qo(i,r)]);

Coefficient (all,i,ENDW_COMM)(all,j,PROD_COMM)(all,r,REG)
REVSHR(i,j,r);
Formula (all,i,ENDW_COMM)(all,j,PROD_COMM)(all,r,REG)
REVSHR(i,j,r) = VFM(i,j,r) / sum(k,PROD_COMM, VFM(i,k,r));
Equation ENDW_PRICE
eq'n generates the composite price for sluggish endowments (HT 50) #   (all,i,ENDWS_COMM)(all,r,REG)   pm(i,r) = sum(k,PROD_COMM, REVSHR(i,k,r) * pmes(i,k,r));

Coefficient (parameter)(all,i,ENDW_COMM)
ETRAE(i)
# elst. of transformation for sluggish primary factor endowments #;
Read
ETRAE from file GTAPPARM header "ETRE";

ETRAE is the elasticity of transformation for sluggish primary factor endowments. It is non-positive, by definition.

Equation ENDW_SUPPLY
eq'n distributes the sluggish endowments across sectors (HT 51)
(all,i,ENDWS_COMM)(all,j,PROD_COMM)(all,r,REG)
qoes(i,j,r) = qo(i,r) - endwslack(i,r) + ETRAE(i) * [pm(i,r) - pmes(i,j,r)];

7-2. Regional Wealth and Equity Income

Variable (all,r,REG)
wqh(r) # wealth of region r #;
Equation EQYHOLDWLTH
eq'n determines shift variable for wealth of the regional hhld
(all,r,REG)
WQHHLD(r) * wqh(r)
= WQHFIRM(r) * wqhf(r) + WQHTRUST(r) * wqht(r);

Equation REGINCEQY
eq'n determines the income from capital in region r
(all, r, REG)
YQ_FIRM(r) * yq_f(r)
= sum(h,ENDWC_COMM, VOA(h,r) * [ps(h,r) + qo(h,r)])
- VDEP(r) * (qk(r) + pcgds(r));

Variable (all,r,REG)
swqh(r) # shift in wealth of region r #;

Equation REGWLTH
This equation determines the change in regional wealth # (all, r, REG) WQHHLD(r) * wqh(r) = WQHFIRM(r) * pcgds(r) + WQHTRUST(r) * pqtrust + 100 * SAVE(r) * time + WQHHLD(r) * swqh(r);

Equation INCHHDLCLEQY
This equation determines the income of the household from local equity # (all, r, REG) yqhf(r) = yq_f(r) + wqhf(r) - wq_f(r);

Equation REGGLBANK
the change in the income of region r from its shares in the global trust # (all, r, REG) yqht(r) = yqt + wqht(r) - wq_t;

Equation TOTINCEQY
This equation determines the change in total income from equity # (all,r,REG) yqh(r) = [YQHFIRM(r) / YQHHLD(r)] * yqhf(r) + [YQHTRUST(r) / YQHHLD(r)] * yqht(r);

7-3. Computation of Regional Income

Coefficient (all,r,REG)
FY(r) # primary factor income in r #;
Formula (all,r,REG)
FY(r) = sum(i,ENDW_COMM, VOM(i,r)) - sum(i,ENDWC_COMM, VOM(i,r));
Variable (all,r,REG)
fincome(r) # factor income #;
Equation FACTORINCOME
factor income #   (all,r,REG)   FY(r) * fincome(r)   = sum(i,ENDW_COMM, VOM(i,r) * [pm(i,r) + qo(i,r)])   - sum(i,ENDWC_COMM, VOM(i,r) * [pm(i,r) + qo(i,r)]);

Variable (change)(all,r,REG)
del_indtaxr(r) # change in ratio of indirect taxes to INCOME in r #;
Equation DINDTAXRATIO
change in ratio of indirect taxes to INCOME in r #   (all,r,REG)   del_indtaxr(r)   = del_taxrpc(r) + del_taxrgc(r) + del_taxriu(r) + del_taxrfu(r)   + del_taxrout(r) + del_taxrexp(r) + del_taxrimp(r);

Variable (change)(all,r,REG)
del_ttaxr(r) # change in ratio of taxes to INCOME in r #;
Equation DTAXRATIO
change in ratio of taxes to INCOME in r #   (all,r,REG)   del_ttaxr(r)   = del_taxrpc(r) + del_taxrgc(r) + del_taxriu(r) + del_taxrfu(r)   + del_taxrout(r) + del_taxrexp(r) + del_taxrimp(r) + del_taxrinc(r);

This variable can be swapped with the commodity- and source-generic consumption tax shift, tp(r), in order to generate a tax replacement scenario, whereby taxes remain a constant share of national income.

Variable (all,r,REG)
incomeslack(r) # slack variable in the expression for regional income #;

This is exogenous, unless the user wishes to fix regional income.

Coefficient (all,r,REG)
INDTAX(r) # indirect tax receipts in r #;
Formula (all,r,REG)
INDTAX(r) = TPC(r) + TGC(r) + TIU(r) + TFU(r) + TOUT(r) + TEX(r) + TIM(r);

Equation REGIONALINCOME
#regional income = sum of income(primary factor+equity)+ indirect tax receipts#   (all,r,REG)   INCOME(r) * y(r)   = FY(r) * fincome(r)   + YQHHLD(r)*yqh(r)   + 100.0 * INCOME(r) * del_indtaxr(r)   + INDTAX(r) * y(r)   + INCOME(r) * incomeslack(r);

7-4. Regional Household Demand System

Variable (all,r,REG)
dpav(r) # average distribution parameter shift, for EV calc. #;
Equation DPARAV
average distribution parameter shift # (all,r,REG) dpav(r) = XSHRPRIV(r) * dppriv(r) + XSHRGOV(r) * dpgov(r) + XSHRSAVE(r) * dpsave(r);

Equation UTILITELASTIC
elasticity of cost of utility wrt utility # (all,r,REG) uelas(r) = XSHRPRIV(r) * uepriv(r) - dpav(r);

Equation PRIVCONSEXP
private consumption expenditure # (all,r,REG) yp(r) - y(r) = -[uepriv(r) - uelas(r)] + dppriv(r);

Equation GOVCONSEXP
government consumption expenditure # (all,r,REG) yg(r) - y(r) = uelas(r) + dpgov(r);

Equation SAVING
saving # (all,r,REG) psave(r) + qsave(r) - y(r) = uelas(r) + dpsave(r);

7-5. Aggregate Utility

Variable (all,r,REG)
p(r) # price index for disposition of income by regional household #;
Equation PRICEINDEXREG
price index for disposition of income by regional household
(all,r,REG)
p(r)
= XSHRPRIV(r) * ppriv(r)
+ XSHRGOV(r) * pgov(r)
+ XSHRSAVE(r) * psave(r);

Variable (all,r,REG)
au(r) # input-neutral shift in utility function #;

Variable (all,r,REG)
dpsum(r) # sum of the distribution parameters #;

Coefficient (all,r,REG)
DPARSUM(r) # sum of distribution parameters #;
Read
DPARSUM from file GTAPDATA header "DPSM";
Update (all,r,REG)
DPARSUM(r) = dpsum(r);

Coefficient (all,r,REG)
UTILELAS(r) # elasticity of cost of utility wrt utility #;
Formula (all,r,REG)
UTILELAS(r)
= [UELASPRIV(r) * XSHRPRIV(r) + XSHRGOV(r) + XSHRSAVE(r)] / DPARSUM(r);

Coefficient (all,r,REG)
DPARPRIV(r) # private consumption distribution parameter #;
Formula (all,r,REG)
DPARPRIV(r) = UELASPRIV(r) * XSHRPRIV(r) / UTILELAS(r);

Coefficient (all,r,REG)
DPARGOV(r) # government consumption distribution parameter #;
Formula (all,r,REG)
DPARGOV(r) = XSHRGOV(r) / UTILELAS(r);

Coefficient (all,r,REG)
DPARSAVE(r) # saving distribution parameter #;
Formula (all,r,REG)
DPARSAVE(r) = XSHRSAVE(r) / UTILELAS(r);

Coefficient (all,r,REG)
UTILPRIV(r) # utility from private consumption #;
Formula (initial)(all,r,REG)
UTILPRIV(r) = 1.0;
Update (all,r,REG)
UTILPRIV(r) = up(r);

Coefficient (all,r,REG)
UTILGOV(r) # utility from government consumption #;
Formula (initial)(all,r,REG)
UTILGOV(r) = 1.0;
Update (all,r,REG)
UTILGOV(r) = ug(r);

Coefficient (all,r,REG)
UTILSAVE(r) # utility from saving #;
Formula (initial)(all,r,REG)
UTILSAVE(r) = 1.0;
Update (change) (all,r,REG)
UTILSAVE(r) = [[qsave(r) - pop(r)] / 100] * UTILSAVE(r);

Variable (all,r,REG)
u(r) # per capita utility from aggregate hhld expend. in region r #;

Equation UTILITY
regional household utility #   (all,r,REG)   u(r)   = au(r)   + DPARPRIV(r) * loge(UTILPRIV(r)) * dppriv(r)   + DPARGOV(r) * loge(UTILGOV(r)) * dpgov(r)   + DPARSAVE(r) * loge(UTILSAVE(r)) * dpsave(r)   + [1.0 / UTILELAS(r)] * [y(r) - pop(r) - p(r)];

Equation DISTPARSUM
sum of the distribution parameters #   (all,r,REG)   DPARSUM(r) * dpsum(r)   = DPARPRIV(r) * dppriv(r)   + DPARGOV(r) * dpgov(r)   + DPARSAVE(r) * dpsave(r);

8. Equilibrium Conditions

8-1. Market Clearing Conditions
8-2. Net Foreign Assets and Liabilities
8-3. Walras' Law
8-4. Trade Balance Constraints

8-1. Market Clearing Conditions

Coefficient (all,i,TRAD_COMM)(all,j,PROD_COMM)(all,r,REG)
SHRDFM(i,j,r) # share of dom. prod. i used by sector j in r at mkt prices #;
Formula (all,i,TRAD_COMM)(all,j,PROD_COMM)(all,r,REG)
SHRDFM(i,j,r) = VDFM(i,j,r) / VDM(i,r);

Coefficient (all,i,TRAD_COMM)(all,r,REG)
SHRDPM(i,r) # share of domestic prod. of i used by private hhlds in r #;
Formula (all,i,TRAD_COMM)(all,r,REG)
SHRDPM(i,r) = VDPM(i,r) / VDM(i,r);

Coefficient (all,i,TRAD_COMM)(all,r,REG)
SHRDGM(i,r) # share of imports of i used by gov't hhlds in r #;
Formula (all,i,TRAD_COMM)(all,r,REG)
SHRDGM(i,r) = VDGM(i,r) / VDM(i,r);

Variable (orig_level=VDM)(all,i,TRAD_COMM)(all,r,REG)
qds(i,r) # domestic sales of commodity i in r #;
Equation MKTCLDOM
eq'n assures market clearing for domestic sales (HT 3) # (all,i,TRAD_COMM)(all,r,REG) qds(i,r) = sum(j,PROD_COMM, SHRDFM(i,j,r) * qfd(i,j,r)) + SHRDPM(i,r) * qpd(i,r) + SHRDGM(i,r) * qgd(i,r);

Coefficient (all,i,TRAD_COMM)(all,r,REG)
SHRDM(i,r) # share of domestic sales of i in r #;
Formula (all,i,TRAD_COMM)(all,r,REG)
SHRDM(i,r) = VDM(i,r) / VOM(i,r);

Coefficient (all,m,MARG_COMM)(all,r,REG)
SHRST(m,r) # share of sales of m to global transport services in r #;
Formula (all,m,MARG_COMM)(all,r,REG)
SHRST(m,r) = VST(m,r) / VOM(m,r);

Coefficient (all,i,TRAD_COMM)(all,r,REG)(all,s,REG)
SHRXMD(i,r,s) # share of export sales of i to s in r #;
Formula (all,i,TRAD_COMM)(all,r,REG)(all,s,REG)
SHRXMD(i,r,s) = VXMD(i,r,s) / VOM(i,r);

Variable (all,i,TRAD_COMM)(all,r,REG)
tradslack(i,r) # slack variable in tradeables market clearing condition #;

This is exogenous unless the user wishes to specify the price of tradeables exogenously, in which case the analysis becomes partial equilibrium and walraslack must be exogenized.

Equation MKTCLTRD_MARG
eq'n assures market clearing for margins commodities (HT 1) # (all,m,MARG_COMM)(all,r,REG) qo(m,r) = SHRDM(m,r) * qds(m,r) + SHRST(m,r) * qst(m,r) + sum(s,REG, SHRXMD(m,r,s) * qxs(m,r,s)) + tradslack(m,r);
Equation MKTCLTRD_NMRG
eq'n assures market clearing for the non-margins commodities (HT 1) # (all,i,NMRG_COMM)(all,r,REG) qo(i,r) = SHRDM(i,r) * qds(i,r) + sum(s,REG, SHRXMD(i,r,s) * qxs(i,r,s)) + tradslack(i,r);

Coefficient (all,i,TRAD_COMM)(all,r,REG)
VIM(i,r) # value of imports of commodity i in r at domestic market prices #;
Formula (all,i,TRAD_COMM)(all,r,REG)
VIM(i,r) = sum(j,PROD_COMM, VIFM(i,j,r)) + VIPM(i,r) + VIGM(i,r);

Coefficient (all,i,TRAD_COMM)(all,j,PROD_COMM)(all,r,REG)
SHRIFM(i,j,r) # share of import i used by sector j in r #;
Formula (all,i,TRAD_COMM)(all,j,PROD_COMM)(all,r,REG)
SHRIFM(i,j,r) = VIFM(i,j,r) / VIM(i,r);

Coefficient (all,i,TRAD_COMM)(all,r,REG)
SHRIPM(i,r) # share of import i used by private hhlds in r #;
Formula (all,i,TRAD_COMM)(all,r,REG)
SHRIPM(i,r) = VIPM(i,r) / VIM(i,r);

Coefficient (all,i,TRAD_COMM)(all,r,REG)
SHRIGM(i,r) # the share of import i used by gov't hhlds in r #;
Formula (all,i,TRAD_COMM)(all,r,REG)
SHRIGM(i,r) = VIGM(i,r) / VIM(i,r);

Equation MKTCLIMP
eq'n assures mkt clearing for imported goods entering each region (HT 2) #   (all,i,TRAD_COMM)(all,r,REG)   qim(i,r)   = sum(j,PROD_COMM, SHRIFM(i,j,r) * qfm(i,j,r))   + SHRIPM(i,r) * qpm(i,r)   + SHRIGM(i,r) * qgm(i,r);

Coefficient (all,i,ENDWM_COMM)(all,j,PROD_COMM)(all,r,REG)
SHREM(i,j,r) # share of mobile endowment i used by sector j at mkt prices #;
Formula (all,i,ENDWM_COMM)(all,j,PROD_COMM)(all,r,REG)
SHREM(i,j,r) = VFM(i,j,r) / VOM(i,r);

Equation MKTCLENDWM
eq'n assures mkt clearing for perfectly mobile endowments in each r (HT 4) #   (all,i,ENDWM_COMM)(all,r,REG)   qo(i,r) = sum(j,PROD_COMM, SHREM(i,j,r) * qfe(i,j,r)) + endwslack(i,r);

This equation assures market clearing for perfectly mobile endowments (HT 4)

Equation MKTCLENDWS
eq'n assures mkt clearing for imperfectly mobile endowments in each r (HT 5) #   (all,i,ENDWS_COMM)(all,j,PROD_COMM)(all,r,REG)   qoes(i,j,r) = qfe(i,j,r);

This equation assures market clearing for sluggish endowments (HT 5)

8-2. Net Foreign Assets and Liabilities

Variable (all,r,REG)
xwq_f(r) # shift variable for value of domestic capital #;

Coefficient (parameter)(all,r,REG)
RIGWQ_F(r) # rigidity of source of funding of enterprises #;
Read
RIGWQ_F from file GTAPPARMK header "RWQF";

Equation EQYHOLDFNDHHD

# eq'n determines the equity holdings of the trust in the household # (all,r,REG) 	
 xwq_f(r) = RIGWQ_F(r) * wqtf(r);

Coefficient (parameter)(all,r,REG)
RIGWQH(r) # rigidity of allocation of wealth by regional houshold #;
Read
RIGWQH from file GTAPPARMK header "RWQH";

Variable (all,r,REG)
xwqh(r) # shift variable for wealth of regional household #;
Variable (all,r,REG)
swqhf(r) # shift equity held by the regional household in domestic firms #;
Equation EQYHOLDHHDLCL
eq'n determines the local equity holdings of the regional hhld #   (all,r,REG)   [RIGWQH(r) + RIGWQ_F(r)] * wqhf(r)   = xwqh(r) + xwq_f(r) + swqhf(r);

Variable (all,r,REG)
swqht(r) # shift equity held by the regional hhld in the global trust #;
Equation EQYHOLDHHDFND
eq'n determines equity holdings of the regional hhld in the global trust
(all,r,REG)
xwqh(r) = RIGWQH(r) * wqht(r) + swqht(r);

8-3. Walras' Law

Variable
wtrustslack # balance sheet shift variable #;
Equation GLOB_BLNC_SHEET
wqt = wq_t + wtrustslack;

Variable
walras_sup # supply in omitted market--global supply of cgds composite #;
Equation WALRAS_S
Extra eq'n computes change in supply in the omitted market. #   GLOBINV * walras_sup   = sum(r,REG, REGINV(r) * [pcgds(r) + qcgds(r)]   - VDEP(r) * [pcgds(r) + qk(r)]);

Variable
walras_dem # demand in the omitted market--global demand for savings #;
Equation WALRAS_D
Extra eq'n computes change in demand in the omitted market. #   GLOBINV * walras_dem = sum(r,REG, SAVE(r) * [psave(r) + qsave(r)]);

Variable
walraslack # slack variable in the omitted market #;

This is endogenous under normal, GE closure. If the GE links are broken, then this must be swapped with the numeraire, thereby forcing global savings to explicitly equal global investment.

Equation WALRAS

# Check Walras' Law. Value of "walraslack" should be zero. (HT 14) # 	
 walras_sup = walras_dem + walraslack;

This equation checks Walras' Law. The value of walraslack should be zero in any GE simulation. (HT 14)

8-4. Trade Balance Constraints

Variable (change)(all,r,REG)
DTBAL(r) # change in trade balance X - M, $ US million #;

Variable (change)(all,r,REG)
SDTBAL(r) # Arbitrary region specific shift variable for TBAL #;

Positive figure indicates increase in exports exceeds imports.

Variable (change)
SDTBALWORLD # Arbitrary region generic shift variable for TBAL #;
Equation TRADEBALANCEF
region specific determination of investment
(all,r,REG)
DTBAL(r) = SDTBAL(r) + SDTBALWORLD;

Appendices

This content is adapted from the dynamic GTAP source code.

For more documentation, refer to:
- Hertel, T.W. and M.E. Tsigas "Structure of the Standard GTAP Model", Chapter 2 in T.W. Hertel (editor) *Global Trade Analysis: Modeling and Applications*, Cambridge University Press, 1997.
- Ianchovichina, E.I. (1998) *International Capital Linkages: Theory and Applications in A Dynamic Computable General Equilibrium Model*.
- Ianchovichina, E.I. and R. McDougall, "Theoretical Structure of Dynamic GTAP", *GTAP Technical Paper No. 17*, Dec. 2000

Appendices

A. Summary Indices
B. Equivalent Variation
C. Welfare Decomposition
D. Terms of Trade Decomposition

A. Summary Indices

The following equations calculate many useful summary statistics. They do not generally affect the equilibrium structure of the model, although they do include the equation for the variable, "qfactsup", which is exogenously given in most of simulations.

A-0. Appendix-Specific Variables and Coefficients
A-1. Factor Price Indices
A-2. Regional Terms of Trade
A-3. GDP Indices (Value, Price and Quantity)
A-4. Aggregate Trade Indices (Value, Price and Quantity)
A-5. Trade Balance Indices

A-0. Appendix-Specific Variables and Coefficients

only used in this Summary Indices appendix

Variable (all,i,TRAD_COMM)(all,s,REG)
vxwfob(i,s) # value of merchandise regional exports, by commodity, FOB #;

Variable (all,i,TRAD_COMM)(all,s,REG)
viwcif(i,s) # value of merchandise regional imports, by commodity, CIF #;

Variable (all,r,REG)
vxwreg(r) # value of merchandise exports, by region #;

Variable (all,r,REG)
viwreg(r) # value of merchandise imports, by region, at world prices #;

Coefficient (all,i,TRAD_COMM)(all,r,REG)
VXW(i,r) # value of exports by comm. i and region r at FOB prices #;
Formula (all,m,MARG_COMM)(all,r,REG)
VXW(m,r) = sum(s,REG, VXWD(m,r,s)) + VST(m,r);
Formula (all,i,NMRG_COMM)(all,r,REG)
VXW(i,r) = sum(s,REG, VXWD(i,r,s));

Coefficient (all,r,REG)
VXWREGION(r) # value of exports by region r at FOB prices #;
Formula (all,r,REG)
VXWREGION(r) = sum(i,TRAD_COMM, VXW(i,r));

Coefficient (all,i,TRAD_COMM)(all,s,REG)
VIW(i,s) # value of commodity imports i into s at CIF prices #;
Formula (all,i,TRAD_COMM)(all,s,REG)
VIW(i,s) = sum(r,REG, VIWS(i,r,s));
Coefficient (all,r,REG)
VIWREGION(r) # value of commodity imports by region r at CIF prices #;
Formula (all,r,REG)
VIWREGION(r) = sum(i,TRAD_COMM, VIW(i,r));

A-1. Factor Price Indices

Variable (orig_level=1.0)(all,i,ENDW_COMM)(all,r,REG)
pfactreal(i,r) # ratio of return to primary factor i to CPI in r #;
Equation REALRETURN
eq'n defines the real rate of return to primary factor i in region r # (all,i,ENDW_COMM)(all,s,REG) pfactreal(i,s) = pm(i,s) - ppriv(s);

This equation defines the real rate of return to primary factor i in region r (new).

Coefficient (all,r,REG)
VENDWREG(r) # value of primary factors, at mkt prices, by region #;
Formula (all,r,REG)
VENDWREG(r) = sum(i,ENDW_COMM, VOM(i,r));

Variable (orig_level=1.0)(all,r,REG)
pfactor(r) # market price index of primary factors, by region #;
Equation PRIMFACTPR

# computes % change in price index of primary factors, by region # (all,r,REG) 	
 VENDWREG(r) * pfactor(r) = sum(i,ENDW_COMM, VOM(i,r) * pm(i,r));

Coefficient
VENDWWLD # value of primary factors, at mkt prices, worldwide #;
Formula
VENDWWLD = sum(r,REG, VENDWREG(r));
Variable (orig_level=1.0)
pfactwld # world price index of primary factors #;
Equation PRIMFACTPRWLD
computes % change in global price index of primary factors #   VENDWWLD * pfactwld = sum(r,REG, VENDWREG(r) * pfactor(r));

Variable (all,i,ENDW_COMM)(all, r, REG)
qfactsup(i,r) # factor supply in region r #;
Variable (all, r, REG)
empl(r) # Employment rate in region r #;
Equation EMPLOY_FACTOR # employment determination condition #
(all,i,ENDW_COMM)(all,r,REG)   qfactsup(i,r) + empl(r) = qo(i,r);

A-2. Regional Terms of Trade

The next three equations correspond to Table 10 of Hertel and Tsigas on Regional Terms of Trade.

Variable (orig_level=1.0)(all,r,REG)
psw(r) # index of prices received for tradeables produced in r #;
Equation REGSUPRICE
estimate change in index of prices received for tradeables i produced in r #   (all,r,REG)   VXWREGION(r) * psw(r)   = sum(i,TRAD_COMM, sum(s,REG, VXWD(i,r,s) * pfob(i,r,s)))   + sum(m,MARG_COMM, VST(m,r) * pm(m,r));

This equation estimates the change in the index of prices received for tradeable products produced in r. (modified from HT 64 to eliminate the investment component)

Variable (orig_level=1.0)(all,r,REG)
pdw(r) # index of prices paid for tradeables used in region r #;
Equation REGDEMPRICE
estimate change in index of prices paid for tradeable products used in r #   (all,r,REG)   VIWREGION(r) * pdw(r)   = sum(i,TRAD_COMM, sum(k,REG, VIWS(i,k,r) * pcif(i,k,r)));

This equation estimates the change in the index of prices paid for tradeable products used in r. (modified from HT 65 to eliminate savings)

Variable (orig_level=1.0)(all,r,REG)
tot(r) # terms of trade for region r: tot(r) = psw(r) - pdw(r) #;
Equation TOTeq
terms of trade equation computed as difference in psw and pdw (HT 66) # (all,r,REG) tot(r) = psw(r) - pdw(r);

A-3. GDP Indices (Value, Price and Quantity)

Coefficient (all,r,REG)
GDP(r) # Gross Domestic Product in region r #;
Formula (all,s,REG)
GDP(s) = sum(i,TRAD_COMM, VPA(i,s)) + sum(i,TRAD_COMM, VGA(i,s)) + sum(k,CGDS_COMM, VOA(k,s)) + sum(i,TRAD_COMM, sum(r,REG, VXWD(i,s,r))) + sum(m,MARG_COMM, VST(m,s)) - sum(i,TRAD_COMM, sum(r,REG, VIWS(i,r,s)));

Gross Domestic Product in region r. Trade is valued at FOB and CIF prices.

Variable (all,r,REG)
vgdp(r) # change in value of GDP #;
Equation VGDP_r
change in value of GDP (HT 70) #   (all,r,REG)   GDP(r) * vgdp(r)   = sum(i,TRAD_COMM, VGA(i,r) * [qg(i,r) + pg(i,r)])   + sum(i,TRAD_COMM, VPA(i,r) * [qp(i,r) + pp(i,r)])   + REGINV(r) * [qcgds(r) + pcgds(r)]   + sum(i,TRAD_COMM, sum(s,REG, VXWD(i,r,s) * [qxs(i,r,s) + pfob(i,r,s)]))   + sum(m,MARG_COMM, VST(m,r) * [qst(m,r) + pm(m,r)])   - sum(i,TRAD_COMM, sum(s,REG, VIWS(i,s,r) * [qxs(i,s,r) + pcif(i,s,r)]));

Variable (orig_level=1.0)(all,r,REG)
pgdp(r) # GDP price index #;
Equation PGDP_r
GDP price index (HT 71) #   (all,r,REG)   GDP(r) * pgdp(r)   = sum(i,TRAD_COMM, VGA(i,r) * pg(i,r))   + sum(i,TRAD_COMM, VPA(i,r) * pp(i,r))   + REGINV(r) * pcgds(r)   + sum(i,TRAD_COMM, sum(s,REG, VXWD(i,r,s) * pfob(i,r,s)))   + sum(m,MARG_COMM, VST(m,r) * pm(m,r))   - sum(i,TRAD_COMM, sum(s,REG, VIWS(i,s,r) * pcif(i,s,r)));

Variable (orig_level=GDP)(all,r,REG)
qgdp(r) # GDP quantity index #;
Equation QGDP_r
GDP quantity index # (all,r,REG) GDP(r) * qgdp(r) = sum(i,TRAD_COMM, VGA(i,r) * qg(i,r)) + sum(i,TRAD_COMM, VPA(i,r) * qp(i,r)) + REGINV(r) * qcgds(r) + sum(i,TRAD_COMM, sum(s,REG, VXWD(i,r,s) * qxs(i,r,s))) + sum(m,MARG_COMM, VST(m,r) * qst(m,r)) - sum(i,TRAD_COMM, sum(s,REG, VIWS(i,s,r) * qxs(i,s,r)));

modified from HT 72 for use with AnalyzeGE

Variable (all,i,PROD_COMM)(all,r,REG)
compvalad(i,r) # composition of value added for good i and region r #;
Equation COMPVALADEQ
track change in composition of value added # (all,i,PROD_COMM)(all,r,REG) compvalad(i,r) = qo(i,r) - qgdp(r);

A-4. Aggregate Trade Indices (Value, Price and Quantity)

Common Variables and Coefficients
Value Indices for Aggregate Trade
Price Indices for Aggregate Trade
Quantity Indices for Aggregate Trade

Common Variables and Coefficients

only used in this Aggregate Trade Indices section

Variable (orig_level=1.0)(all,i,TRAD_COMM)(all,r,REG)
pxw(i,r) # aggregate exports price index of i from region r #;

Coefficient (all,i,TRAD_COMM)
VXWCOMMOD(i) # value of world exports by commodity i at FOB prices #;
Formula (all,i,TRAD_COMM)
VXWCOMMOD(i) = sum(r,REG, VXW(i,r));

Coefficient
VXWLD # value of commodity exports, FOB, globally #;
Formula
VXWLD = sum(r,REG, VXWREGION(r));

Coefficient (all,i,TRAD_COMM)
VIWCOMMOD(i) # global value of commodity imports, CIF, by commodity #;
Formula (all,i,TRAD_COMM)
VIWCOMMOD(i) = sum(r,REG, VIW(i,r));

Coefficient (all,i,TRAD_COMM)(all,r,REG)
PW_PM(i,r) # ratio of world to domestic prices #;
Formula (all,i,TRAD_COMM)(all,r,REG)
PW_PM(i,r) = sum(s,REG, VXWD(i,r,s)) / sum(s,REG, VXMD(i,r,s));
Coefficient (all,i,TRAD_COMM)(all,r,REG)
VOW(i,r) # value of output in r at FOB including transportation services #;
Formula (all,m,MARG_COMM)(all,r,REG)
VOW(m,r) = VDM(m,r) * PW_PM(m,r) + sum(s,REG, VXWD(m,r,s)) + VST(m,r);
Formula (all,i,NMRG_COMM)(all,r,REG)
VOW(i,r) = VDM(i,r) * PW_PM(i,r) + sum(s,REG, VXWD(i,r,s));

Coefficient (all,i,TRAD_COMM)
VWOW(i) # value of world supply at world prices for i #;
Formula (all,i,TRAD_COMM)
VWOW(i) = sum(r,REG, VOW(i,r));

Coefficient (all,i,TRAD_COMM)
VWOU(i) # value of world output of i at user prices #;
Formula (all,i,TRAD_COMM)
VWOU(i) = sum(s,REG, [VPA(i,s) + VGA(i,s)] + sum(j,PROD_COMM, VFA(i,j,s)));

Value Indices for Aggregate Trade

Equation VREGEX_ir_MARG
the change in FOB value of exports of m from r #   (all,m,MARG_COMM)(all,r,REG)   VXW(m,r) * vxwfob(m,r)   = sum(s,REG, VXWD(m,r,s) * [qxs(m,r,s) + pfob(m,r,s)])   + VST(m,r) * [qst(m,r) + pm(m,r)];
Equation VREGEX_ir_NMRG
the change in FOB value of exports of commodity i from r (HT 73) #   (all,i,NMRG_COMM)(all,r,REG)   VXW(i,r) * vxwfob(i,r)   = sum(s,REG, VXWD(i,r,s) * [qxs(i,r,s) + pfob(i,r,s)]);

Equation VREGEX_r
computes % change in value of merchandise exports, by region (HT 75) #   (all,r,REG)   VXWREGION(r) * vxwreg(r) = sum(i,TRAD_COMM, VXW(i,r) * vxwfob(i,r));

Variable (all,i,TRAD_COMM)
vxwcom(i) # value of global merchandise exports by commodity #;
Equation VWLDEX_i
computes % change in FOB value of global exports, by commodity (HT 77) #   (all,i,TRAD_COMM)   VXWCOMMOD(i) * vxwcom(i) = sum(r,REG, VXW(i,r) * vxwfob(i,r));

Variable
vxwwld # value of world trade #;
Equation VWLDEX
computes % change in value of global exports (HT 79) #   VXWLD * vxwwld = sum(r,REG, VXWREGION(r) * vxwreg(r));

Equation VREGIM_is
the change in CIF value of imports of commodity i into s (HT 74) # (all,i,TRAD_COMM)(all,s,REG) VIW(i,s) * viwcif(i,s) = sum(r,REG, VIWS(i,r,s) * [pcif(i,r,s) + qxs(i,r,s)]);

Equation VREGIM_s
computes % change in value of imports, CIF basis, by region (HT 76) # (all,s,REG) VIWREGION(s) * viwreg(s) = sum(i,TRAD_COMM, VIW(i,s) * viwcif(i,s));

Variable (all,i,TRAD_COMM)
viwcom(i) # value of global merchandise imports i, at world prices #;

Equation VWLDIM_i
computes % change in value of global imports, by commodity (HT 78) # (all,i,TRAD_COMM) VIWCOMMOD(i) * viwcom(i) = sum(s,REG, VIW(i,s) * viwcif(i,s));

Variable (all,i,TRAD_COMM)
valuew(i) # value of world supply of good i #;
Equation VWLDOUT
change in value of world output of comm. i at FOB prices (HT 80) #   (all,i,TRAD_COMM)   VWOW(i) * valuew(i) = sum(r,REG, VOW(i,r) * [pxw(i,r) + qo(i,r)]);

Variable (all,i,TRAD_COMM)
valuewu(i) # value of world supply of good i at user prices #;
Equation VWLDOUTUSE
change in value of world output of commodity i at user prices #   (all,i,TRAD_COMM)   VWOU(i) * valuewu(i)   = sum(s,REG, VPA(i,s) * [pp(i,s) + qp(i,s)]   + VGA(i,s) * [pg(i,s) + qg(i,s)]   + sum(j,PROD_COMM, VFA(i,j,s) * [pf(i,j,s) + qf(i,j,s)]));

Price Indices for Aggregate Trade

Equation PREGEX_ir_MARG
change in FOB price index of exports of m from r #   (all,m,MARG_COMM)(all,r,REG)   VXW(m,r) * pxw(m,r)   = sum(s,REG, VXWD(m,r,s) * pfob(m,r,s)) + VST(m,r) * pm(m,r);

Equation PREGEX_ir_NMRG
change in FOB price index of exports of commodity i from r (HT 81) #   (all,i,NMRG_COMM)(all,r,REG)   VXW(i,r) * pxw(i,r) = sum(s,REG, VXWD(i,r,s) * pfob(i,r,s));

Variable (orig_level=1.0)(all,r,REG)
pxwreg(r) # price index of merchandise exports, by region #;
Equation PREGEX_r
computes % change in price index of exports, by region (HT 83) #   (all,r,REG)   VXWREGION(r) * pxwreg(r) = sum(i,TRAD_COMM, VXW(i,r) * pxw(i,r));

Variable (orig_level=1.0)(all,i,TRAD_COMM)
pxwcom(i) # price index of global merchandise exports by commodity #;
Equation PWLDEX_i
computes % change in price index of exports, by commodity (HT 85) #   (all,i,TRAD_COMM)   VXWCOMMOD(i) * pxwcom(i) = sum(r,REG, VXW(i,r) * pxw(i,r));

Variable (orig_level=1.0)
pxwwld # price index of world trade #;
Equation PWLDEX
computes % change in price index of global exports (HT 87) #   VXWLD * pxwwld = sum(r,REG, VXWREGION(r) * pxwreg(r));

Variable (all,i,TRAD_COMM)(all,r,REG)
piw(i,r) # world price of composite import i in region r #;
Equation PREGIM_is
change in CIF price index of imports of commodity i into s (HT 82) #   (all,i,TRAD_COMM)(all,s,REG)   VIW(i,s) * piw(i,s) = sum(r,REG, VIWS(i,r,s) * pcif(i,r,s));

Variable (orig_level=1.0)(all,r,REG)
piwreg(r) # price index of merchandise imports, by region #;
Equation PREGIM_s
computes % change in price index of imports, by region (HT 84) #   (all,s,REG)   VIWREGION(s) * piwreg(s) = sum(i,TRAD_COMM, VIW(i,s) * piw(i,s));

Variable (orig_level=1.0)(all,i,TRAD_COMM)
piwcom(i) # price index of global merchandise imports by commodity #;
Equation PWLDIM_i
computes % change in price index of imports, by commodity (HT 86) # (all,i,TRAD_COMM) VIWCOMMOD(i) * piwcom(i) = sum(s,REG, VIW(i,s) * piw(i,s));

Variable (all,i,TRAD_COMM)
pw(i) # world price index for total good i supplies #;
Equation PWLDOUT
change in index of world prices, FOB, for total production of i (HT 88) # (all,i,TRAD_COMM) VWOW(i) * pw(i) = sum(r,REG, VOW(i,r) * pxw(i,r));

Variable (orig_level=1.0)(all,i,TRAD_COMM)
pwu(i) # world price index for total good i supplies at user prices #;
Equation PWLDUSE
change in index of user prices for deflating world production of i # (all,i,TRAD_COMM) VWOU(i) * pwu(i) = sum(s,REG, VPA(i,s) * pp(i,s) + VGA(i,s) * pg(i,s) + sum(j,PROD_COMM, VFA(i,j,s) * pf(i,j,s)));

Quantity Indices for Aggregate Trade

Variable (orig_level=VXW)(all,i,TRAD_COMM)(all,r,REG)
qxw(i,r) # aggregate exports of i from region r, FOB weights #;
Equation QREGEX_ir_MARG
change in volume of exports of margin commodity m from r # (all,m,MARG_COMM)(all,r,REG) VXW(m,r) * qxw(m,r) = sum(s,REG, VXWD(m,r,s) * qxs(m,r,s)) + VST(m,r) * qst(m,r);
Equation QREGEX_ir_NMRG
change in volume of exports of non-margin commodity i from r # (all,i,NMRG_COMM)(all,r,REG) VXW(i,r) * qxw(i,r) = sum(s,REG, VXWD(i,r,s) * qxs(i,r,s));

Variable (orig_level=VXWREGION)(all,r,REG)
qxwreg(r) # volume of merchandise exports, by region #;
Equation QREGEX_r
computes % change in quantity index of exports, by region # (all,r,REG) VXWREGION(r) * qxwreg(r) = sum(i,TRAD_COMM, VXW(i,r) * qxw(i,r));

Variable (orig_level=VXWCOMMOD)(all,i,TRAD_COMM)
qxwcom(i) # volume of global merchandise exports by commodity #;
Equation QWLDEX_i
computes % change in quantity index of exports, by commodity # (all,i,TRAD_COMM) VXWCOMMOD(i) * qxwcom(i) = sum(r,REG, VXW(i,r) * qxw(i,r));

Variable (orig_level=VXWLD)
qxwwld # volume of world trade #;
Equation QWLDEX
computes % change in quantity index of global exports # VXWLD * qxwwld = sum(r,REG, VXWREGION(r) * qxwreg(r));

Variable (all,i,TRAD_COMM)(all,s,REG)
qiw(i,s) # aggregate imports of i into region s, CIF weights #;
Equation QREGIM_is
change in volume of imports of commodity i into s # (all,i,TRAD_COMM)(all,s,REG) VIW(i,s) * qiw(i,s) = sum(r,REG, VIWS(i,r,s) * qxs(i,r,s));

Variable (orig_level=VIWREGION)(all,r,REG)
qiwreg(r) # volume of merchandise imports, by region #;
Equation QREGIM_s
computes % change in quantity index of imports, by region # (all,s,REG) VIWREGION(s) * qiwreg(s) = sum(i,TRAD_COMM, VIW(i,s) * qiw(i,s));

Variable (orig_level=VIWCOMMOD)(all,i,TRAD_COMM)
qiwcom(i) # volume of global merchandise imports by commodity #;
Equation QWLDIM_i
computes % change in quantity index of imports, by commodity # (all,i,TRAD_COMM) VIWCOMMOD(i) * qiwcom(i) = sum(s,REG, VIW(i,s) * qiw(i,s));

Variable (all,i,TRAD_COMM)
qow(i) # quantity index for world supply of good i #;
Equation QWLDOUT
change in index of world production of i # (all,i,TRAD_COMM) VWOW(i) * qow(i) = sum(r,REG, VOW(i,r) * qo(i,r));

Variable (orig_level=VWOU)(all,i,TRAD_COMM)
qowu(i) # quantity index for world supply of good i at user prices #;
Equation QWLDOUTU
change in index of world production of i evaluated at user prices # (all,i,TRAD_COMM) VWOU(i) * qowu(i) = sum(s,REG, VPA(i,s) * qp(i,s) + VGA(i,s) * qg(i,s) + sum(j,PROD_COMM, VFA(i,j,s) * qf(i,j,s)));

A-5. Trade Balance Indices

Variable (change)(all,i,TRAD_COMM)(all,r,REG)
DTBALi(i,r) # change in trade balance by i and by r, $ US million #;

A positive value indicates that the change in exports exceeds the change in imports.

Equation TRADEBAL_i
computes change in trade balance by commodity and by region (HT 97) #   (all,i,TRAD_COMM)(all,r,REG)   DTBALi(i,r)   = [VXW(i,r) / 100] * vxwfob(i,r) - [VIW(i,r) / 100] * viwcif(i,r);

```
Equation TRADEBALANCE
# computes change in trade balance (X - M), by region (HT 98) #
(all,r,REG)
DTBAL(r)
 = [VXWREGION(r) / 100] * vxwreg(r) - [VIWREGION(r) / 100] *
viwreg(r);
```

In order to maintain homogeneity in the presence of a fixed trade balance, it is useful to have a nominal variable which this is measured against. The next equation provides this, and we recommend users fix DTBALR instead of fixing DTBAL in future simulations. The strategy is the same one used above for taxes.

```
Coefficient (all,r,REG)
TBAL(r) # trade balance for region r #;
Formula (all,r,REG)
TBAL(r) = VXWREGION(r) - VIWREGION(r);
```

```
Variable (change)(all,r,REG)
DTBALR(r) # change in ratio of trade balance to regional income #;
Equation DTBALRATIO
# change in ratio of trade balance to regional income #
(all,r,REG)
100 * INCOME(r) * DTBALR(r) = 100 * DTBAL(r) - TBAL(r) * y(r);
```

B. Equivalent Variation

B-0. Appendix-Specific Variables and Coefficients
B-1. Government Consumption Shadow Demand System
B-2. Private Consumption Shadow Demand System
B-3. Regional Household Shadow Demand System
B-4. Equivalent Variation

This appendix calculates equivalent variation "EV" and world equivalent variation, "WEV", by determining the income "yev" that would be required to achieve the current actual utility level "u" in a shadow demand system in which prices are fixed.

Section B-2 calculates the utility elasticity of private consumption expenditure, "ueprivev", within a shadow demand system for private consumption, for use in section B-3. B-3 calculates private consumption expenditure "ypev" for use in B-2, and regional income "yev" for use in B-4, within a shadow demand system for the regional household. B-4 calculates "EV" and "WEV".

B-0. Appendix-Specific Variables and Coefficients

only used in this Equivalent Variation section of the Summary Indices appendix

Variable (all,r,REG)
uelasev(r)   # elasticity of cost of utility wrt utility, for EV calc. #;

Variable (all,r,REG)
ueprivev(r)   # utility elasticity of private consn expenditure, for EV calc. #;

Variable (all,r,REG)
ugev(r) # per capita utility from gov't expend., for EV calc. #;

Variable (all,r,REG)
upev(r) # per capita utility from private expend., for EV calc. #;

Variable (all,r,REG)
qsaveev(r) # total quantity of savings demanded, for EV calc. #;

Variable (all,r,REG)
yev(r) # regional household income in region r, for EV calc. #;

Variable (all,r,REG)
ypev(r)   # private consumption expenditure in region r, for EV calc. #;

Variable (all,r,REG)
ygev(r)   # government consumption expenditure in region r, for EV calc. #;

Coefficient (all,r,REG)
INCOMEEV(r) # regional income, for EV calc. #;
Formula (initial) (all,r,REG)
INCOMEEV(r) = INCOME(r);
Update (all,r,REG)
INCOMEEV(r) = yev(r);

Coefficient (all,r,REG)
UTILPRIVEV(r) # utility from private consumption, for EV calcs #;
Formula (initial) (all,r,REG)
UTILPRIVEV(r) = UTILPRIV(r);
Update (all,r,REG)
UTILPRIVEV(r) = upev(r);

Coefficient (all,r,REG)
UTILGOVEV(r) # utility from private consumption, for EV calcs #;
Formula (initial) (all,r,REG)
UTILGOVEV(r) = UTILGOV(r);
Update (all,r,REG)
UTILGOVEV(r) = ugev(r);

Coefficient (all,r,REG)
UTILSAVEEV(r) # utility from private consumption, for EV calcs #;
Formula (initial) (all,r,REG)
UTILSAVEEV(r) = UTILSAVE(r);
Update (change) (all,r,REG)
UTILSAVEEV(r) = [[qsaveev(r) - pop(r)] / 100] * UTILSAVEEV(r);

B-1. Government Consumption Shadow Demand System

Equation GOVUSHD
utility from government consumption in r
(all,r,REG)
ygev(r) - pop(r) = ugev(r);

B-2. Private Consumption Shadow Demand System

Variable (all,i,TRAD_COMM)(all,r,REG)
qpev(i,r)
# private hhld demand for commodity i in region r, for EV calc. #;

Coefficient (all,i,TRAD_COMM)(all,r,REG)
VPAEV(i,r)
# private hhld expend. on i in r valued at agent's prices, for EV calc. #;
Formula (initial) (all,i,TRAD_COMM)(all,r,REG)
VPAEV(i,r) = VPA(i,r);
Update (all,i,TRAD_COMM)(all,r,REG)
VPAEV(i,r) = qpev(i,r);

Coefficient (all,r,REG)
VPAREGEV(r) # private consumption expenditure in region r, for EV calc. #;
Formula (all,r,REG)
VPAREGEV(r) = sum(i,TRAD_COMM, VPAEV(i,r));

VPAREGEV should agree with PRIVEXPEV.

Coefficient (all,i,TRAD_COMM)(all,r,REG)
CONSHREV(i,r)
# share of private hhld consn devoted to good i in r, for EV calc. #;
Formula (all,i,TRAD_COMM)(all,r,REG)
CONSHREV(i,r) = VPAEV(i,r) / VPAREGEV(r);

Coefficient (all,i,TRAD_COMM)(all,r,REG)
EYEV(i,r)
# expend. elast. of private hhld demand for i in r, for EV calc. #;
Formula (all,i,TRAD_COMM)(all,r,REG)
EYEV(i,r) = [1.0 / sum(n,TRAD_COMM, CONSHREV(n,r) * INCPAR(n,r))] * [INCPAR(i,r) * [1.0 - ALPHA(i,r)] + sum(n,TRAD_COMM, CONSHREV(n,r) * INCPAR(n,r) * ALPHA(n,r))] + ALPHA(i,r) - sum(n,TRAD_COMM, CONSHREV(n,r) * ALPHA(n,r));

Equation PRIVDMNDSEV
private hhld demands for composite commodities, for EV calc.
(all,i,TRAD_COMM)(all,r,REG)
qpev(i,r) - pop(r) = EYEV(i,r) * [ypev(r) - pop(r)];

Prices are held constant for the EV calculation and so do not appear here.

Coefficient (all,r,REG)
UELASPRIVEV(r)
# elast. of cost wrt utility from private consn, for EV calc. #;
Formula (all,r,REG)
UELASPRIVEV(r) = sum(i,TRAD_COMM, CONSHREV(i,r) * INCPAR(i,r));

Equation PRIVATEUEV
computation of utility from private consumption in r (HT 45)
(all,r,REG)
ypev(r) - pop(r) = UELASPRIVEV(r) * upev(r);

Coefficient (all,i,TRAD_COMM)(all,r,REG)
XWCONSHREV(i,r)
# expansion-parameter-weighted consumption share, for EV calc. #;
Formula (all,i,TRAD_COMM)(all,r,REG)
XWCONSHREV(i,r) = CONSHREV(i,r) * INCPAR(i,r) / UELASPRIVEV(r);

Equation UTILELASPRIVEV
elasticity of cost wrt utility from private consn, for EV calc.
(all,r,REG)
ueprivev(r)
= sum(i,TRAD_COMM, XWCONSHREV(i,r) * [qpev(i,r) - ypev(r)]);

Prices are held constant for the EV calculation and so do not appear here.

B-3. Regional Household Shadow Demand System

Variable (all,r,REG)
ysaveev(r) # NET savings expenditure, for EV calc. #;

Coefficient (all,r,REG)
PRIVEXPEV(r)
# private consumption expenditure in region r, for EV calc. #;
Formula (initial) (all,r,REG)
PRIVEXPEV(r) = PRIVEXP(r);
Update (all,r,REG)
PRIVEXPEV(r) = ypev(r);

PRIVEXPEV should agree with VPAREGEV.

Coefficient (all,r,REG)
GOVEXPEV(r)
# government consumption expenditure in region r, for EV calc. #;
Formula (initial) (all,r,REG)
GOVEXPEV(r) = GOVEXP(r);
Update (all,r,REG)
GOVEXPEV(r) = ygev(r);

Coefficient (all,r,REG)
SAVEEV(r)
# saving in region r, for EV calc. #;
Formula (initial) (all,r,REG)
SAVEEV(r) = SAVE(r);
Update (all,r,REG)
SAVEEV(r) = ysaveev(r);

Coefficient (all,r,REG)
XSHRPRIVEV(r)
# private expenditure share in regional income, for EV calc. #;
Formula (all,r,REG)
XSHRPRIVEV(r) = PRIVEXPEV(r) / INCOMEEV(r);

Coefficient (all,r,REG)
XSHRGOVEV(r)
# government expenditure share in regional income, for EV calc. #;
Formula (all,r,REG)
XSHRGOVEV(r) = GOVEXPEV(r) / INCOMEEV(r);

Coefficient (all,r,REG)
XSHRSAVEEV(r) # saving share in regional income, for EV calc. #;
Formula (all,r,REG)
XSHRSAVEEV(r) = SAVEEV(r) / INCOMEEV(r);

Variable (all,r,REG)
dpavev(r) # average distribution parameter shift, for EV calc. #;
Equation DPARAVEV
average distribution parameter shift, for EV calc. # (all,r,REG) dpavev(r) = XSHRPRIVEV(r) * dppriv(r) + XSHRGOVEV(r) * dpgov(r) + XSHRSAVEEV(r) * dpsave(r);

Equation UTILITELASTICEV
elasticity of cost of utility wrt utility, for EV calc. # (all,r,REG) uelasev(r) = XSHRPRIV(r) * ueprivev(r) - dpavev(r);

Equation PCONSEXPEV
private consumption expenditure, for EV calc. # (all,r,REG) ypev(r) - yev(r) = -[ueprivev(r) - uelasev(r)] + dppriv(r);

Equation GOVCONSEXPEV
government consumption expenditure # (all,r,REG) ygev(r) - yev(r) = uelasev(r) + dpgov(r);

Equation SAVINGEV
saving # (all,r,REG) ysaveev(r) - yev(r) = uelasev(r) + dpsave(r);

Equation SAVEUEV
saving
(all,r,REG)
qsaveev(r) = ysaveev(r);

Note that because psave doesn't change, qsaveev moves with ysaveev.

Coefficient (all,r,REG)
UTILELASEV(r)
# elasticity of cost of utility wrt utility, for EV calc. #;
Formula (initial) (all,r,REG)
UTILELASEV(r) = UTILELAS(r);
Update (all,r,REG)
UTILELASEV(r) = uelasev(r);

Equation INCOME_EQUIV
equivalent income
(all,r,REG)
u(r)
= au(r)
+ DPARPRIV(r) * loge(UTILPRIVEV(r)) * dppriv(r)
+ DPARGOV(r) * loge(UTILGOVEV(r)) * dpgov(r)
+ DPARSAVE(r) * loge(UTILSAVEEV(r)) * dpsave(r)
+ [1.0 / UTILELASEV(r)] * [yev(r) - pop(r)];

B-4. Equivalent Variation

Variable (change)(all,r,REG)
EV(r) # equivalent variation, $ US million #;
Equation EVREG
regional EV (HT 67)
(all,r,REG)
EV(r) = [INCOMEEV(r) / 100] * yev(r);

Variable (change)
WEV # equivalent variation for the world #;
Equation EVWLD
# EV for the world (HT 68) #
WEV = sum(r,REG, EV(r));

C. Welfare Decomposition

See GTAP Technical Paper No. 5 for derivation and interpretation.

Coefficient (all,r,REG)
EVSCALFACT(r) # equivalent variation scaling factor #;
Formula (all,r,REG)
EVSCALFACT(r) = [UTILELASEV(r) / UTILELAS(r)] * [INCOMEEV(r) / INCOME(r)];

Coefficient (all,m,MARG_COMM)(all,s,REG)
VTMD(m,s) # aggregate value of svce m in shipments to s #;
Formula (all,m,MARG_COMM)(all,s,REG)
VTMD(m,s) = sum(i,TRAD_COMM, sum(r,REG, VTMFSD(m,i,r,s)));

Variable (linear,change)(all,r,REG)
EV_ALT(r) # regional EV computed in alternative way #;
Equation EV_DECOMPOSITION

decomposition of Equivalent Variation
(all,r,REG)
EV_ALT(r)
= -[0.01 * UTILELASEV(r) * INCOMEEV(r)]
* [DPARPRIV(r) * loge(UTILPRIVEV(r) / UTILPRIV(r)) * dppriv(r)
+ DPARGOV(r) * loge(UTILGOVEV(r) / UTILGOV(r)) * dpgov(r)
+ DPARSAVE(r) * loge(UTILSAVEEV(r) / UTILSAVE(r)) * dpsave(r)]
+ [0.01 * EVSCALFACT(r)]
* [sum(i,NSAV_COMM, PTAX(i,r) * [qo(i,r) - pop(r)])
+ sum(i,ENDW_COMM, sum(j,PROD_COMM,
ETAX(i,j,r) * [qfe(i,j,r) - pop(r)]))
+ sum(i,TRAD_COMM, sum(j,PROD_COMM,
IFTAX(i,j,r) * [qfm(i,j,r) - pop(r)]))
+ sum(i,TRAD_COMM, sum(j,PROD_COMM,
DFTAX(i,j,r) * [qfd(i,j,r) - pop(r)]))
+ sum(i,TRAD_COMM, IPTAX(i,r) * [qpm(i,r) - pop(r)])
+ sum(i,TRAD_COMM, DPTAX(i,r) * [qpd(i,r) - pop(r)])
+ sum(i,TRAD_COMM, IGTAX(i,r) * [qgm(i,r) - pop(r)])
+ sum(i,TRAD_COMM, DGTAX(i,r) * [qgd(i,r) - pop(r)])
+ sum(i,TRAD_COMM, sum(s,REG, XTAXD(i,r,s) * [qxs(i,r,s) - pop(r)]))
+ sum(i,TRAD_COMM, sum(s,REG, MTAX(i,s,r) * [qxs(i,s,r) - pop(r)]))
+ sum(i,ENDW_COMM, VOA(i,r) * [qo(i,r) - pop(r)])
- VDEP(r) * [qk(r) - pop(r)]
+ YQHTRUST(r) * (yqht(r) - pop(r)) - YQTFIRM(r) * (yqtf(r) - pop(r))
+ sum(i,PROD_COMM, VOA(i,r) * ao(i,r))
+ sum(j,PROD_COMM, VVA(j,r) * ava(j,r))
+ sum(j,PROD_COMM, sum(i,ENDW_COMM, VFA(i,j,r) * afe(i,j,r)))
+ sum(j,PROD_COMM, sum(i,TRAD_COMM, VFA(i,j,r) * af(i,j,r)))
+ sum(m,MARG_COMM, sum(i,TRAD_COMM, sum(s,REG,
VTMFSD(m,i,s,r) * atmfsd(m,i,s,r))))
+ sum(i,TRAD_COMM, sum(s,REG, VIMS(i,s,r) * ams(i,s,r)))
+ sum(i,TRAD_COMM, sum(s,REG, VXWD(i,r,s) * pfob(i,r,s)))
+ sum(m,MARG_COMM, VST(m,r) * pm(m,r))
+ NETINV(r) * pcgds(r)
- sum(i,TRAD_COMM, sum(s,REG, VXWD(i,s,r) * pfob(i,s,r)))
- sum(m,MARG_COMM, VTMD(m,r) * pt(m))
- SAVE(r) * psave(r)]
+ 0.01 * INCOMEEV(r) * pop(r);

Variable (linear,change)
WEV_ALT # expression for WEV computed in alternative way #;
Equation WORLDEV
# Equivalent Variation for the world #
WEV_ALT = sum(r,REG, EV_ALT(r));

Variable (linear,change) (all,r,REG)
CNTdpar(r) # contribution to EV of change in distribution parameters #;
Equation CNT_WEV_dpar
(all,r,REG) CNTdpar(r) = -0.01 * UTILELASEV(r) * INCOMEEV(r) * [DPARPRIV(r) * loge(UTILPRIVEV(r) / UTILPRIV(r)) * dppriv(r) + DPARGOV(r) * loge(UTILGOVEV(r) / UTILGOV(r)) * dpgov(r) + DPARSAVE(r) * loge(UTILSAVEEV(r) / UTILSAVE(r)) * dpsave(r)];

Variable (linear,change) (all,r,REG)
CNTpopr(r) # contribution to EV in region r of change in population #;
Equation CONT_EV_pop
(all,r,REG) CNTpopr(r) = 0.01 * INCOMEEV(r) * pop(r);

Variable (linear,change) (all,r,REG)
CNTqor(r) # contribution to EV in region r of output changes #;
Equation CONT_EV_qor
(all,r,REG) CNTqor(r) = sum(i,NSAV_COMM, 0.01 * EVSCALFACT(r) * PTAX(i,r) * [qo(i,r) - pop(r)]);

Variable (linear,change) (all,i,NSAV_COMM)(all,r,REG)
CNTqoir(i,r) # contribution to EV of changes in output of NSAV_COMM i in reg. r #;
Equation CONT_EV_qoir
(all,i,NSAV_COMM)(all,r,REG) CNTqoir(i,r) = PTAX(i,r) * [0.01 * EVSCALFACT(r)] * [qo(i,r) - pop(r)];

Variable (linear,change) (all,r,REG)
CNTqfer(r)   # cont. to EV of changes in use of all ENDW_COMM in all ind. in reg. r #;
Equation CONT_EV_qfer
(all,r,REG)   CNTqfer(r)   = sum(i,ENDW_COMM, sum(j,PROD_COMM,   ETAX(i,j,r) * [0.01 * EVSCALFACT(r)] * [qfe(i,j,r) - pop(r)]));

Variable (linear,change) (all,i,ENDW_COMM)(all,r,REG)
CNTqfeir(i,r)   # contribution to EV of changes in use of ENDW_COMM i in all ind. in r #;
Equation CONT_EV_qfeir
(all,i,ENDW_COMM)(all,r,REG)   CNTqfeir(i,r)   = sum(j,PROD_COMM,   ETAX(i,j,r) * [0.01 * EVSCALFACT(r)] * [qfe(i,j,r) - pop(r)]);

Variable (linear,change)   (all,i,ENDW_COMM)(all,j,PROD_COMM)(all,r,REG)
CNTqfeijr(i,j,r)   # cont. to EV of changes in use of ENDW_COMM i in ind. j of reg. r #;
Equation CONT_EV_qfeijr
(all,i,ENDW_COMM)(all,j,PROD_COMM)(all,r,REG)   CNTqfeijr(i,j,r)   = ETAX(i,j,r) * [0.01 * EVSCALFACT(r)] * [qfe(i,j,r) - pop(r)];

Variable (linear,change) (all,r,REG)
CNTqfmr(r)   # cont. to EV of changes in use of imported int. in all ind. in reg. r #;
Equation CONT_EV_qfmr
(all,r,REG)   CNTqfmr(r)   = sum(i,TRAD_COMM, sum(j,PROD_COMM,   IFTAX(i,j,r) * [0.01 * EVSCALFACT(r)] * [qfm(i,j,r) - pop(r)]));

Variable (linear,change) (all,i,TRAD_COMM)(all,r,REG)
CNTqfmir(i,r)
# cont. to EV of changes in use of imported int. i in all ind. in r #;
Equation CONT_EV_qfmir
(all,i,TRAD_COMM)(all,r,REG)   CNTqfmir(i,r)   = sum(j,PROD_COMM,   IFTAX(i,j,r) * [0.01 * EVSCALFACT(r)] * [qfm(i,j,r) - pop(r)]);

Variable (linear,change)   (all,i,TRAD_COMM)(all,j,PROD_COMM)(all,r,REG)
CNTqfmijr(i,j,r)
# cont. to EV of changes in use of imported int. i in ind. j of reg. r #;
Equation CONT_EV_qfmijr
(all,i,TRAD_COMM)(all,j,PROD_COMM)(all,r,REG)   CNTqfmijr(i,j,r)   = IFTAX(i,j,r) * [0.01 * EVSCALFACT(r)] * [qfm(i,j,r) - pop(r)];

Variable (linear,change) (all,r,REG)
CNTqfdr(r)
# cont. to EV of changes in use of domestic int. in all ind. in reg. r #;
Equation CONT_EV_qfdr
(all,r,REG)   CNTqfdr(r)   = sum(i,TRAD_COMM, sum(j,PROD_COMM,   DFTAX(i,j,r) * [0.01 * EVSCALFACT(r)] * [qfd(i,j,r) - pop(r)]));

Variable (linear,change) (all,i,TRAD_COMM)(all,r,REG)
CNTqfdir(i,r)
# contribution to EV of changes in use of domestic i in all ind. in r #;
Equation CONT_EV_qfdir
(all,i,TRAD_COMM)(all,r,REG)   CNTqfdir(i,r)   = sum(j,PROD_COMM,   DFTAX(i,j,r) * [0.01 * EVSCALFACT(r)] * [qfd(i,j,r) - pop(r)]);

Variable (linear,change) (all,i,TRAD_COMM)(all,j,PROD_COMM)(all,r,REG)
CNTqfdijr(i,j,r)
# cont. to EV of changes in use of domestic int. i in ind. j of reg. r #;
Equation CONT_EV_qfdijr
(all,i,TRAD_COMM)(all,j,PROD_COMM)(all,r,REG) CNTqfdijr(i,j,r) = DFTAX(i,j,r) * [0.01 * EVSCALFACT(r)] * [qfd(i,j,r) - pop(r)];

Variable (linear,change) (all,r,REG)
CNTqpmr(r)
# contribution to EV of changes in consumption of imported goods in r #;
Equation CONT_EV_qpmr
(all,r,REG) CNTqpmr(r) = sum(i,TRAD_COMM, IPTAX(i,r) * [0.01 * EVSCALFACT(r)] * [qpm(i,r) - pop(r)]);

Variable (linear,change) (all,i,TRAD_COMM)(all,r,REG)
CNTqpmir(i,r)
# cont. to EV of changes in consumption of imported good i in reg. r #;
Equation CONT_EV_qpmir
(all,i,TRAD_COMM)(all,r,REG) CNTqpmir(i,r) = IPTAX(i,r) * [0.01 * EVSCALFACT(r)] * [qpm(i,r) - pop(r)];

Variable (linear,change) (all,r,REG)
CNTqpdr(r)
# contribution to EV of changes in consumption of domestic goods in r #;
Equation CONT_EV_qpdr
(all,r,REG) CNTqpdr(r) = sum(i,TRAD_COMM, DPTAX(i,r) * [0.01 * EVSCALFACT(r)] * [qpd(i,r) - pop(r)]);

Variable (linear,change) (all,i,TRAD_COMM)(all,r,REG)
CNTqpdir(i,r)
# cont. to EV of changes in consumption of domestic good i in reg. r #;
Equation CONT_EV_qpdir
(all,i,TRAD_COMM)(all,r,REG) CNTqpdir(i,r) = DPTAX(i,r) * [0.01 * EVSCALFACT(r)] * [qpd(i,r) - pop(r)];

Variable (linear,change) (all,r,REG)
CNTqgmr(r)
# cont. to EV of changes in gov't consumption of imports in reg. r #;
Equation CONT_EV_qgmr
(all,r,REG) CNTqgmr(r) = sum(i,TRAD_COMM, IGTAX(i,r) * [0.01 * EVSCALFACT(r)] * [qgm(i,r) - pop(r)]);

Variable (linear,change) (all,i,TRAD_COMM)(all,r,REG)
CNTqgmir(i,r)
# cont. to EV of changes in gov't consumption of import i in reg. r #;
Equation CONT_EV_qgmir
(all,i,TRAD_COMM)(all,r,REG) CNTqgmir(i,r) = IGTAX(i,r) * [0.01 * EVSCALFACT(r)] * [qgm(i,r) - pop(r)];

Variable (linear,change) (all,r,REG)
CNTqgdr(r) # cont. to EV of changes in gov't consumption of domestics in reg. r #;
Equation CONT_EV_qgdr
(all,r,REG) CNTqgdr(r) = sum(i,TRAD_COMM, DGTAX(i,r) * [0.01 * EVSCALFACT(r)] * [qgd(i,r) - pop(r)]);

Variable (linear,change) (all,i,TRAD_COMM)(all,r,REG)
CNTqgdir(i,r) # cont. to EV of changes in gov't consumption of domestic i in reg. r #;
Equation CONT_EV_qgdir
(all,i,TRAD_COMM)(all,r,REG) CNTqgdir(i,r) = DGTAX(i,r) * [0.01 * EVSCALFACT(r)] * [qgd(i,r) - pop(r)];

Variable (linear,change) (all,r,REG)
CNTqxsr(r) # cont. to EV of changes in exports of all goods from SRCE r to all DEST #;
Equation CONT_EV_qxsr
(all,r,REG) CNTqxsr(r) = sum(i,TRAD_COMM, sum(s,REG, XTAXD(i,r,s) * [0.01 * EVSCALFACT(r)] * [qxs(i,r,s) - pop(r)]));

Variable (linear,change) (all,i,TRAD_COMM)(all,r,REG)(all,s,REG)
CNTqxsirs(i,r,s)
# cont. to EV of changes in exports of i from SRCE r to DEST s #;
Equation CONT_EV_qxsirs
(all,i,TRAD_COMM)(all,r,REG)(all,s,REG)
CNTqxsirs(i,r,s)
= XTAXD(i,r,s) * [0.01 * EVSCALFACT(r)] * [qxs(i,r,s) - pop(r)];

Variable (linear,change) (all,r,REG)
CNTqimr(r)
# cont. to EV of changes in imports of all goods from all SRCE to DEST r #;
Equation CONT_EV_qimr
(all,r,REG)
CNTqimr(r)
= sum(i,TRAD_COMM, sum(s,REG, MTAX(i,s,r) * [0.01 * EVSCALFACT(r)] * [qxs(i,s,r) - pop(r)]));

Variable (linear,change) (all,i,TRAD_COMM)(all,s,REG)(all,r,REG)
CNTqimisr(i,s,r)
# cont. to EV of changes in imports of i from SRCE s to DEST r #;
Equation CONT_EV_qimisr
(all,i,TRAD_COMM)(all,s,REG)(all,r,REG)
CNTqimisr(i,s,r)
= MTAX(i,s,r) * [0.01 * EVSCALFACT(r)] * [qxs(i,s,r) - pop(r)];

Variable (linear,change) (all,r,REG)
CNTalleffr(r) # total contribution to regional EV of allocative effects #;
Equation CONT_EV_alleffr
(all,r,REG) CNTalleffr(r) = [0.01 * EVSCALFACT(r)] * [sum(i,NSAV_COMM, PTAX(i,r) * [qo(i,r) - pop(r)]) + sum(i,ENDW_COMM, sum(j,PROD_COMM, ETAX(i,j,r) * [qfe(i,j,r) - pop(r)])) + sum(i,TRAD_COMM, sum(j,PROD_COMM, IFTAX(i,j,r) * [qfm(i,j,r) - pop(r)])) + sum(i,TRAD_COMM, sum(j,PROD_COMM, DFTAX(i,j,r) * [qfd(i,j,r) - pop(r)])) + sum(i,TRAD_COMM, IPTAX(i,r) * [qpm(i,r) - pop(r)]) + sum(i,TRAD_COMM, DPTAX(i,r) * [qpd(i,r) - pop(r)]) + sum(i,TRAD_COMM, IGTAX(i,r) * [qgm(i,r) - pop(r)]) + sum(i,TRAD_COMM, DGTAX(i,r) * [qgd(i,r) - pop(r)]) + sum(i,TRAD_COMM, sum(s,REG, XTAXD(i,r,s) * [qxs(i,r,s) - pop(r)])) + sum(i,TRAD_COMM, sum(s,REG, MTAX(i,s,r) * [qxs(i,s,r) - pop(r)]))];

Variable (linear,change) (all,i,DEMD_COMM)(all,r,REG)
CNTalleffir(i,r)   # total contribution to regional EV of allocative effects #;
Equation CONT_EV_alleffir_E
(all,i,ENDW_COMM)(all,r,REG)   CNTalleffir(i,r)   = [0.01 * EVSCALFACT(r)]   * [PTAX(i,r) * [qo(i,r) - pop(r)]   + sum(j,PROD_COMM, ETAX(i,j,r) * [qfe(i,j,r) - pop(r)])];
Equation CONT_EV_alleffir_T
(all,i,TRAD_COMM)(all,r,REG)   CNTalleffir(i,r)   = [0.01 * EVSCALFACT(r)]   * [PTAX(i,r) * [qo(i,r) - pop(r)]   + sum(j,PROD_COMM, IFTAX(i,j,r) * [qfm(i,j,r) - pop(r)])   + sum(j,PROD_COMM, DFTAX(i,j,r) * [qfd(i,j,r) - pop(r)])   + IPTAX(i,r) * [qpm(i,r) - pop(r)]   + DPTAX(i,r) * [qpd(i,r) - pop(r)]   + IGTAX(i,r) * [qgm(i,r) - pop(r)]   + DGTAX(i,r) * [qgd(i,r) - pop(r)]   + sum(s,REG, XTAXD(i,r,s) * [qxs(i,r,s) - pop(r)])   + sum(s,REG, MTAX(i,s,r) * [qxs(i,s,r) - pop(r)])];

Variable (linear,change) (all,r,REG)
CNTtotr(r)
# contribution to regional EV of changes in its terms of trade #;
Equation CONT_EV_totr
(all,r,REG) CNTtotr(r) = [0.01 * EVSCALFACT(r)] * [sum(i,TRAD_COMM, sum(s,REG, VXWD(i,r,s) * [pfob(i,r,s)])) + sum(m,MARG_COMM, VST(m,r) * [pm(m,r)]) - sum(i,TRAD_COMM, sum(s,REG, VXWD(i,s,r) * [pfob(i,s,r)])) - sum(m,MARG_COMM, VTMD(m,r) * [pt(m)])];

Variable (linear,change) (all,r,REG)
CNTcgdsr(r) # contribution to regional EV of changes in the price of cgds #;
Equation CNT_EV_cgdsr
(all,r,REG) CNTcgdsr(r) = [0.01 * EVSCALFACT(r)] * [NETINV(r) * [pcgds(r)] - SAVE(r) * [psave(r)]];

VARIABLE (LINEAR,CHANGE) (all,r,REG) CNTendwnar(r)
Contribution to regional EV of changes in all non-accum endowments # ;
EQUATION CONT_EV_endwnar (all,r,REG)
CNTendwnar(r) = [0.01 * EVSCALFACT(r)] * [sum(i,ENDWNA_COMM, VOA(i,r) * [qo(i,r) - pop(r)])] ;

VARIABLE (LINEAR,CHANGE) (all,i,ENDWNA_COMM)(all,r,REG) CNTendwnair(i,r)
Contribution to regional EV of changes non-accum endowment i # ;
EQUATION CONT_EV_endwnair (all,i,ENDWNA_COMM)(all,r,REG)
CNTendwnair(i,r) = [0.01 * EVSCALFACT(r)] * [VOA(i,r) * [qo(i,r) - pop(r)]] ;

VARIABLE (LINEAR,CHANGE) (all,r,REG) CNTendwcr(r)
Contribution to regional EV of changes in Capital used in region r # ;
EQUATION CONT_EV_endwcr (all,r,REG)
CNTendwcr(r) = [0.01 * EVSCALFACT(r)] * [sum(i,ENDWC_COMM, VOA(i,r) * [qk(r) - pop(r)]) - VDEP(r) * [qk(r) - pop(r)]] ;

VARIABLE (LINEAR,CHANGE) (all,r,REG) CNTfeqyr(r)
Contrib'n to regional EV of changes in financial equity owned by region r # ;
EQUATION CONT_EV_feqyr (all,r,REG)
CNTfeqyr(r) = [0.01 * EVSCALFACT(r)] * [sum(i,ENDWC_COMM, VOA(i,r) * [qk(r) - pop(r)]) - VDEP(r) * [qk(r) - pop(r)] + YQHTRUST(r) * [yqht(r) - pop(r)] - YQTFIRM(r) * [yqtf(r) - pop(r)]] ;

VARIABLE (LINEAR,CHANGE) (all,r,REG) CNTeqytr(r)
Contrib'n to regional EV due to foreign ownership in and by region r # ;
EQUATION CONT_EV_eqytr (all,r,REG)
CNTeqytr(r) = [0.01 * EVSCALFACT(r)] * [YQHTRUST(r) * [yqht(r) - pop(r)] - YQTFIRM(r) * [yqtf(r) - pop(r)]] ;

VARIABLE (LINEAR,CHANGE) (all,r,REG) CNTqht(r)
Contribution to regional EV of changes in income from foreign equity # ;
EQUATION CONT_EV_qht (all,r,REG)
CNTqht(r) = [0.01 * EVSCALFACT(r)] * [(YQHTRUST(r) * [yqht(r) - pop(r)])] ;

VARIABLE (LINEAR,CHANGE) (all,r,REG) CNTqtf(r)
Contrb to reg EV of chgs in ownership of domestic capital # ;
EQUATION CONT_EV_dror (all,r,REG)
CNTqtf(r) = [0.01 * EVSCALFACT(r)] * [- YQTFIRM(r) * [yqtf(r) - pop(r)]] ;

Variable (linear,change) (all,r,REG)
CNTtechr(r) # contribution to regional EV of all technical change #;
Equation CONT_EV_techr
(all,r,REG) CNTtechr(r) = [0.01 * EVSCALFACT(r)] * [sum(i,PROD_COMM, VOA(i,r) * ao(i,r)) + sum(j,PROD_COMM, sum(i,ENDW_COMM, VFA(i,j,r) * afe(i,j,r))) + sum(j,PROD_COMM, VVA(j,r) * ava(j,r)) + sum(j,PROD_COMM, sum(i,TRAD_COMM, VFA(i,j,r) * af(i,j,r))) + sum(m,MARG_COMM, sum(i,TRAD_COMM, sum(s,REG, VTMFSD(m,i,s,r) * atmfsd(m,i,s,r)))) + sum(i,TRAD_COMM, sum(s,REG, VIMS(i,s,r) * ams(i,s,r)))];

Variable (linear,change) (all,r,REG)
CNTtech_aor(r) # contribution to regional EV of output augmenting technical change #;
Equation CONT_EV_tech_aor
(all,r,REG) CNTtech_aor(r) = [0.01 * EVSCALFACT(r)] * sum(i,PROD_COMM, VOA(i,r) * ao(i,r));

Variable (linear,change) (all,i,PROD_COMM)(all,r,REG)
CNTtech_aoir(i,r)   # cont. to regional EV of output augmenting tech change in TRAD_COMM i #;
Equation CONT_EV_tech_aoir
(all,i,PROD_COMM)(all,r,REG)   CNTtech_aoir(i,r) = [0.01 * EVSCALFACT(r)] * VOA(i,r) * ao(i,r);

Variable (linear,change) (all,r,REG)
CNTtech_afer(r)   # contribution to regional EV of primary factor augmenting tech change #;
Equation CONT_EV_tech_afer
(all,r,REG)   CNTtech_afer(r)   = [0.01 * EVSCALFACT(r)]   * sum(j,PROD_COMM, sum(i,ENDW_COMM, VFA(i,j,r) * afe(i,j,r)));

Variable (linear,change) (all,i,ENDW_COMM)(all,j,PROD_COMM)(all,r,REG)
CNTtech_afeijr(i,j,r)   # cont. to EV of primary factor i augmenting tech change in sector j #;
Equation CONT_EV_tech_afeijr
(all,i,ENDW_COMM)(all,j,PROD_COMM)(all,r,REG)   CNTtech_afeijr(i,j,r) = [0.01 * EVSCALFACT(r)] * VFA(i,j,r) * afe(i,j,r);

Variable (linear,change) (all,r,REG)
CNTtech_avar(r)   # contribution to regional EV of value added augmenting tech change #;
Equation CONT_EV_tech_avar
(all,r,REG)   CNTtech_avar(r)   = [0.01 * EVSCALFACT(r)] * sum(j,PROD_COMM, VVA(j,r) * ava(j,r));

Variable (linear,change) (all,j,PROD_COMM)(all,r,REG)
CNTtech_avajr(j,r)   # cont. to EV of value added augmenting tech change in sector j #;
Equation CONT_EV_tech_avajr
(all,j,PROD_COMM)(all,r,REG)   CNTtech_avajr(j,r) = [0.01 * EVSCALFACT(r)] * VVA(j,r) * ava(j,r);

Variable (linear,change) (all,r,REG)
CNTtech_afr(r)   # cont. to regional EV of comp. int. input augmenting tech change #;
Equation CONT_EV_tech_afr
(all,r,REG)   CNTtech_afr(r)   = [0.01 * EVSCALFACT(r)]   * sum(j,PROD_COMM, sum(i,TRAD_COMM,   [VIFA(i,j,r) + VDFA(i,j,r)] * af(i,j,r)));

Variable (linear,change) (all,i,TRAD_COMM)(all,j,PROD_COMM)(all,r,REG)
CNTtech_afijr(i,j,r) # cont. to EV of composite i input augmenting tech change in sector j #;
Equation CONT_EV_tech_afijr
(all,i,TRAD_COMM)(all,j,PROD_COMM)(all,r,REG) CNTtech_afijr(i,j,r) = [0.01 * EVSCALFACT(r)] * [VIFA(i,j,r) + VDFA(i,j,r)] * af(i,j,r);

Variable (linear,change) (all,r,REG)
CNTtech_atrr(r) # contribution to regional EV of technical change in transportation #;
Equation CONT_EV_tech_atrr
(all,r,REG) CNTtech_atrr(r) = [0.01 * EVSCALFACT(r)] * sum(m,MARG_COMM, sum(i,TRAD_COMM, sum(s,REG, VTMFSD(m,i,s,r) * atmfsd(m,i,s,r))));

Variable (linear,change)(all,m,MARG_COMM)(all,i,TRAD_COMM)(all,r,REG)(all,s,REG)
CNTtech_afmfdsd(m,i,r,s) # cont. to EV of tech change in transportation efficiency #;
Equation CONT_EV_tech_afmfdsd
(all,m,MARG_COMM)(all,i,TRAD_COMM)(all,r,REG)(all,s,REG) CNTtech_afmfdsd(m,i,r,s) = [0.01 * EVSCALFACT(s)] * VTMFSD(m,i,r,s) * atmfsd(m,i,r,s);

Variable (linear,change)(all,r,REG)
CNTtech_amsr(r) # cont. to EV of bilateral import-augmenting tech change #;
Equation CONT_EV_tech_amsr
(all,r,REG) CNTtech_amsr(r) = [0.01 * EVSCALFACT(r)] * sum(i,TRAD_COMM, sum(s,REG, VIMS(i,s,r) * ams(i,s,r)));

Variable (linear,change) (all,i,TRAD_COMM)(all,r,REG)(all,s,REG)
CNTtech_amsirs(i,r,s) # cont. to EV of bilateral import augmenting tech change for TRAD_COMM i #;
Equation CONT_EV_tech_amsirs
(all,i,TRAD_COMM)(all,r,REG)(all,s,REG) CNTtech_amsirs(i,r,s) = [0.01 * EVSCALFACT(r)] * VIMS(i,r,s) * ams(i,r,s);

Variable (linear,change) (all,r,REG)
CNTkbr(r)
# cont. to EV of changes to beg. capital stock and depreciation #;
Equation CONT_EV_kbr
(all,r,REG)
CNTkbr(r) = - [0.01 * EVSCALFACT(r)] * VDEP(r) * [qk(r) - pop(r)];

D. Terms of Trade Decomposition

Computations for decomposition the terms-of-trade effect
Reference: Robert McDougall, SALTER No. 12 Working Paper

Coefficient (all,i,TRAD_COMM)(all,r,REG)
SX_IR(i,r) # share of good i in total exports from r #;
Formula (all,m,MARG_COMM)(all,r,REG)
SX_IR(m,r) = [sum(s,REG, VXWD(m,r,s)) + VST(m,r)] / [sum(k,TRAD_COMM, sum(s,REG, VXWD(k,r,s))) + sum(l,MARG_COMM, VST(l,r))];
Formula (all,i,NMRG_COMM)(all,r,REG)
SX_IR(i,r) = sum(s,REG, VXWD(i,r,s)) / [sum(k,TRAD_COMM, sum(s,REG, VXWD(k,r,s))) + sum(l,MARG_COMM, VST(l,r))];

Coefficient (all,m,MARG_COMM)(all,s,REG)
VTICOMM(m,s) # margin usage of m in getting imports to region s #;
Formula (all,m,MARG_COMM)(all,s,REG)
VTICOMM(m,s) = sum(i,TRAD_COMM, sum(r,REG, VTMFSD(m,i,r,s)));

Coefficient (all,i,TRAD_COMM)(all,s,REG)
VIWDIRALL(i,s) # imports of i into s, with direct allocation of margins #;
Formula (all,m,MARG_COMM)(all,s,REG)
VIWDIRALL(m,s) = sum(r,REG, VXWD(m,r,s)) + VTICOMM(m,s);
Formula (all,i,NMRG_COMM)(all,s,REG)
VIWDIRALL(i,s) = sum(r,REG, VXWD(i,r,s));

Coefficient (all,s,REG)
VIWDATOT(s) # total imports into s, calculated on direct allocation basis #;
Formula (all,s,REG)
VIWDATOT(s) = sum(i,TRAD_COMM, VIWDIRALL(i,s));

Coefficient (all,i,TRAD_COMM)(all,r,REG)
SM_IR(i,r) # share of good i in total imports into r #;
Formula (all,i,TRAD_COMM)(all,s,REG)
SM_IR(i,s) = VIWDIRALL(i,s) / VIWDATOT(s);

Zerodivide (zero_by_zero) default 0;
Coefficient (all,i,TRAD_COMM)(all,r,REG)(all,s,REG)
SX_IRS(i,r,s) # share of exports of good i from region r to s #;
Formula (all,m,MARG_COMM)(all,r,REG)(all,s,REG)
SX_IRS(m,r,s) = VXWD(m,r,s) / [sum(k,REG, VXWD(m,r,k)) + VST(m,r)];
Formula (all,i,NMRG_COMM)(all,r,REG)(all,s,REG)
SX_IRS(i,r,s) = VXWD(i,r,s) / sum(k,REG, VXWD(i,r,k));
Zerodivide (zero_by_zero) off;

Coefficient (all,m,MARG_COMM)(all,r,REG)
SXT_IR(m,r) # share of margins in exports of good i from region r #;
Formula (all,m,MARG_COMM)(all,r,REG)
SXT_IR(m,r) = VST(m,r) / [sum(k,REG, VXWD(m,r,k)) + VST(m,r)];

Variable (all,i,TRAD_COMM)(all,r,REG)
px_ir(i,r) # export price index for good i and region r #;
Equation EXPPRICE_MARG
price index for total exports of m from r # (all,m,MARG_COMM)(all,r,REG) px_ir(m,r) = sum(s,REG, SX_IRS(m,r,s) * pfob(m,r,s)) + SXT_IR(m,r) * pm(m,r);

Equation EXPPRICE_NMRG
price index for total exports of i from r # (all,i,NMRG_COMM)(all,r,REG) px_ir(i,r) = sum(s,REG, SX_IRS(i,r,s) * pfob(i,r,s));

Coefficient (all,i,TRAD_COMM)(all,r,REG)(all,s,REG)
SM_IRS(i,r,s) # share of imports of good i into s from r at FOB prices #;
Formula (all,i,TRAD_COMM)(all,r,REG)(all,s,REG)
SM_IRS(i,r,s) = VXWD(i,r,s) / VIWDIRALL(i,s);

Coefficient (all,m,MARG_COMM)(all,r,REG)
SMT_IR(m,r) # share of transport cost in imports of margin commodity #;
Formula (all,m,MARG_COMM)(all,s,REG)
SMT_IR(m,s) = VTICOMM(m,s) / VIWDIRALL(m,s);

Variable (all,i,TRAD_COMM)(all,r,REG)
pm_ir(i,r) # imports price index for good i and region r #;
Equation IMPPRICE1_MARG
price index for total imports of m in s -- margins commodities # (all,m,MARG_COMM)(all,s,REG) pm_ir(m,s) = sum(r,REG, SM_IRS(m,r,s) * pfob(m,r,s)) + SMT_IR(m,s) * pt(m);

Equation IMPPRICE1_NMRG
price index for total imports of i in s -- non-margins commodities # (all,i,NMRG_COMM)(all,s,REG) pm_ir(i,s) = sum(r,REG, SM_IRS(i,r,s) * pfob(i,r,s));

Coefficient (all,i,TRAD_COMM)(all,r,REG)
SW_IR(i,r) # share of region r exports in world total for good i #;
Formula (all,m,MARG_COMM)(all,r,REG)
SW_IR(m,r) = [sum(s,REG, VXWD(m,r,s)) + VST(m,r)] / sum(k,REG, sum(s,REG, VXWD(m,k,s)) + VST(m,k));
Formula (all,i,NMRG_COMM)(all,r,REG)
SW_IR(i,r) = sum(s,REG, VXWD(i,r,s)) / sum(k,REG, sum(s,REG, VXWD(i,k,s)));

Variable (all,i,TRAD_COMM)
px_i(i) # world export price index for commodity i #;
Equation WRLDPRICEi
world export price index for good i #   (all,i,TRAD_COMM)   px_i(i) = sum(r,REG, SW_IR(i,r) * px_ir(i,r));

Coefficient (all,i,TRAD_COMM)
SW_I(i) # share of exports of i in world total #;
Formula (all,m,MARG_COMM)
SW_I(m)   = sum(r,REG, sum(s,REG, VXWD(m,r,s)) + VST(m,r))   / [sum(k,TRAD_COMM, sum(r,REG, sum(s,REG, VXWD(k,r,s))))   + sum(l,MARG_COMM, sum(r,REG, VST(l,r)))];
Formula (all,i,NMRG_COMM)
SW_I(i)   = sum(r,REG, sum(s,REG, VXWD(i,r,s)))   / [sum(k,TRAD_COMM, sum(r,REG, sum(s,REG, VXWD(k,r,s))))   + sum(l,MARG_COMM, sum(r,REG, VST(l,r)))];

Variable
px_ # world export price index for all commodities #;
Equation WRLDPRICE

# world export price index for all goods # 	
 px_ = sum(i,TRAD_COMM, SW_I(i) * px_i(i));

Variable (all,i,TRAD_COMM)(all,r,REG)
c1_ir(i,r) # contribution of world price, px_i, to ToT #;
Equation c1_irEQ
contribution of world export price index of good i to ToT for region r #   (all,i,TRAD_COMM)(all,r,REG)   c1_ir(i,r) = [SX_IR(i,r) - SM_IR(i,r)] * [px_i(i) - px_];

Variable (all,i,TRAD_COMM)(all,r,REG)
c2_ir(i,r) # contribution of regional export price, px_ir, to ToT #;
Equation c2_irEQ
contribution of regional export price of good i for region r #   (all,i,TRAD_COMM)(all,r,REG)   c2_ir(i,r) = SX_IR(i,r) * [px_ir(i,r) - px_i(i)];

Variable (all,i,TRAD_COMM)(all,r,REG)
c3_ir(i,r) # contribution of regional import price, pm_ir, to ToT #;
Equation c3_irEQ
contribution of imports price index of good i for region r #   (all,i,TRAD_COMM)(all,r,REG)   c3_ir(i,r) = SM_IR(i,r) * [pm_ir(i,r) - px_i(i)];

Variable (all,r,REG)
c1_r(r) # contribution of world prices for all goods to ToT #;
Equation c1_rEQ
contribution of world price indexes of all goods to ToT for r #   (all,r,REG)   c1_r(r) = sum(i,TRAD_COMM, c1_ir(i,r));

Variable (all,r,REG)
c2_r(r) # contribution of regional export prices to ToT #;
Equation c2_rEQ
contribution of regional exports prices to ToT for r #   (all,r,REG)   c2_r(r) = sum(i,TRAD_COMM, c2_ir(i,r));

Variable (all,r,REG)
c3_r(r) # contribution of regional import prices to ToT #;
Equation c3_rEQ
contribution of regional import prices to ToT for r #   (all,r,REG)   c3_r(r) = sum(i,TRAD_COMM, c3_ir(i,r));

Variable (all,r,REG)
tot2(r) # trade terms for region r, computed from components #;
Equation TOT2eq
trade terms for region r, computed from components #   (all,r,REG)   tot2(r) = c1_r(r) + c2_r(r) - c3_r(r);

ABOUT THE AUTHOR

Turkay Yildiz received his Ph.D. from the Institute of Marine Sciences and Technology, Dokuz Eylul University, Izmir, Turkey. He received his Master's Degree in Logistics Management from Izmir University of Economics. He has a number of peer reviewed publications and conference presentations at various countries in such fields as transportation, logistics and supply chains. He also has various levels of expertise in the applications of Information Technology.

www.ingramcontent.com/pod-product-compliance
Lightning Source LLC
Chambersburg PA
CBHW070557300426
44113CB00010B/1294